DEPARTMENT OF THE ENVIRONMENT

Survey of Land for Mineral Working in England, 1988

VOLUME 1
REPORT ON SURVEY RESULTS

LONDON: HMSO

© Crown copyright 1991
First published 1991

ISBN 0 11 752365 8

Acknowledgements

Front cover pictures:
Sand and gravel working and restored land at Shepperton, West London. Photos DOE with permission of Greenham Construction Materials Ltd.

Aerial view of Tunstead and Old Moor Limestone Quarries in the Peak District National Park. Photo reproduced with permission from H White, Peak Park Special Planning Board.

CONTENTS

	Page
Executive Summary	xvi
1. Introduction	1
2. Purpose and Scope of the Survey	1
3. Organisation of the Survey	1
4. The response	2
5. Comparison with 1982 and 1974 Surveys	2
6. Presentation of Results	3

PART 1   NATIONAL COMMENTARY

7. Surface Mineral Workings — 7

Introduction — 7
Permissions which include reclamation conditions — 8
Permissions with no provisions for reclamation — 13
Areas where provisions for reclamation were unfulfilled — 15
Regional comparisons — 16
Areas dependant on fill — 16
Areas included in the 1988 Derelict Land Survey — 16

8. Surface Disposal of Mineral Working Deposits — 20

Introduction — 20
Permissions which include reclamation conditions — 22
Permissions with no provisions for reclamation — 22
Areas where provisions for reclamation were unfulfilled — 29
Regional comparisons — 29
Areas included in the 1988 Derelict Land Survey — 30

9. Land covered by Permissions which include Aftercare Conditions — 31

Introduction — 31
Surface mineral workings — 32
Surface disposal of mineral working deposits — 33
Choice of afteruse — 34
Regional comparisons — 37

10. Permissions for Underground Mining — 38

11. Reclamation of Mineral Workings — 41

Introduction — 41
Area of land reclaimed — 41
Choice of afteruse — 45
Agriculture — 45
Forestry — 45
Amenity — 46
Other uses — 46
Land reclaimed satisfactorily — 48

		Page
	Land reclaimed unsatisfactorily	48
	Regional comparisons	48
	Arrangements through which reclamation was achieved	49

12. Number of Sites — 52

Introduction — 52
Surface mineral workings — 52
Surface disposal of mineral working deposits — 54
Underground mining — 57
Regional comparisons — 57

13. Removal of Material from Mineral Working Deposits — 59

14. Reviews under the Town and Country Planning Act 1971 — 61

PART 2 REGIONAL COMMENTARY

15. Northern Region — 65

Introduction — 65
Surface mineral workings — 65
Surface disposal of mineral working deposits — 68
Underground mining — 70
Permissions which include aftercare conditions — 71
Reworking of Mineral Working Deposits — 71
Reclamation — 74
Number of sites — 74
Land included in the Derelict Land Survey — 76
Reviews under the Town and Country Planning Act 1971 — 77

16. Yorkshire and Humberside — 78

Introduction — 78
Surface mineral workings — 78
Surface disposal of mineral working deposits — 81
Underground mining — 82
Permissions which include aftercare conditions — 84
Reworking of Mineral Working Deposits — 84
Reclamation — 84
Number of sites — 88
Land included in the Derelict Land Survey — 89
Reviews under the Town and Country Planning Act 1971 — 90

17. East Midlands — 91

Introduction — 91
Surface mineral workings — 91
Surface disposal of mineral working deposits — 94
Underground mining — 95
Permissions which include aftercare conditions — 95
Reworking of Mineral Working Deposits — 97
Reclamation — 97
Number of sites — 99
Land included in the Derelict Land Survey — 99
Reviews under the Town and Country Planning Act 1971 — 103

		Page
18.	**East Anglia**	104
	Introduction	104
	Surface mineral workings	104
	Surface disposal of mineral working deposits	105
	Underground mining	105
	Permissions which include aftercare conditions	107
	Reworking of Mineral Working Deposits	107
	Reclamation	107
	Number of sites	109
	Land included in the Derelict Land Survey	112
	Reviews under the Town and Country Planning Act 1971	112
19.	**South East**	113
	Introduction	113
	Surface mineral workings	113
	Surface disposal of mineral working deposits	117
	Underground mining	117
	Permissions which include aftercare conditions	119
	Reworking of Mineral Working Deposits	121
	Reclamation	121
	Number of sites	121
	Land included in the Derelict Land Survey	124
	Reviews under the Town and Country Planning Act 1971	124
20.	**South West**	126
	Introduction	126
	Surface mineral workings	126
	Surface disposal of mineral working deposits	128
	Underground mining	131
	Permissions which include aftercare conditions	131
	Reworking of Mineral Working Deposits	133
	Reclamation	133
	Number of sites	137
	Land included in the Derelict Land Survey	137
	Reviews under the Town and Country Planning Act 1971	138
21.	**West Midlands**	139
	Introduction	139
	Surface mineral workings	139
	Surface disposal of mineral working deposits	140
	Underground mining	142
	Permissions which include aftercare conditions	145
	Reworking of Mineral Working Deposits	145
	Reclamation	145
	Number of sites	149
	Land included in the Derelict Land Survey	150
	Reviews under the Town and Country Planning Act 1971	150
22.	**North West**	151
	Introduction	151
	Surface mineral workings	151

	Page
Surface disposal of mineral working deposits	154
Underground mining	154
Permissions which include aftercare conditions	157
Reworking of Mineral Working Deposits	157
Reclamation	157
Number of sites	161
Land included in the Derelict Land Survey	161
Reviews under the Town and Country Planning Act 1971	162

Appendix 163

Sample survey form	163
General Guidance Notes	164
Glossary	167
Tables for Completion	169

INDEX TO TABLES

TABLE	TITLE
7.1	Area of land permitted for surface mineral workings in England in 1982 and 1988.
7.2	Area of surface workings with satisfactory reclamation conditions in 1982 and 1988 indicating the percentage of the total permitted area.
7.3	Area of surface mineral workings with no provisions for reclamation in 1982 and 1988 indicating the percentage of the total permitted area.
8.1	Area of land permitted for spoil disposal in England in 1982 and 1988.
8.2	Area of spoil disposal permissions with satisfactory reclamation conditions in 1982 and 1988 indicating the percentage of the total permitted area.
8.3	Area of spoil disposal permissions with no provisions for reclamation in 1982 and 1988 indicating the percentage of the total permitted area.
9.1	Area and number of permissions which include aftercare conditions for surface mineral workings and spoil disposal.
10.1	Extent of underground mining permissions by mineral type in England in 1982 and 1988.
11.1	Total area of land reclaimed and average annual rate of reclamation in England between 1974–1982 and 1982–1988. All minerals.
12.1	Number of sites affected by mineral working in England in 1982 and 1988.
12.2	Number of surface mineral workings in England in 1982 and 1988 indicating change in number and percentage change - all minerals.
12.3	Number of sites for the surface disposal of mineral working deposits in England in 1982 and 1988 indicating change in number and percentage change - all minerals.
12.4	Number of underground workings in England in 1982 and 1988 indicating change in number and percentage change - all minerals.
13.1	Number of sites and total area permitted for the removal of material from mineral working deposits.
14.1	Mineral planning authorities which had begun their Minerals Review by 1 April 1988, and progress made on reviews.
14.2	Number of sites and unimplemented permissions reviewed by mineral planning authorities by 1 April 1988, and progress made on reviews.
15.1	Total permitted area for surface mineral workings in the Northern region indicating the percentage regional contribution to the England total.
15.2	Total permitted area for the surface disposal of mineral working deposits in the Northern region indicating the percentage regional contribution to the England total.
15.3	Total area of underground permissions in the Northern region indicating the percentage regional contribution to the England total.
15.4	Number of sites for mineral workings in the Northern region in 1988.
16.1	Total permitted area for surface mineral workings in Yorkshire and Humberside, indicating the percentage regional contribution to the England total.

TABLE TITLE

16.2	Total permitted area for the surface disposal of mineral working deposits in Yorkshire and Humberside, indicating the percentage regional contribution to the England total.
16.3	Total area of underground permissions in Yorkshire and Humberside, indicating the percentage regional contribution to the England total.
16.4	Number of sites for mineral workings in Yorkshire and Humberside in 1988.
17.1	Total permitted area for surface mineral workings in the East Midlands region, indicating the percentage regional contribution to the England total.
17.2	Total permitted area for the surface disposal of mineral working deposits in the East Midlands region, indicating the percentage regional contribution to the England total.
17.3	Total area of underground permissions in the East Midlands region, indicating the percentage regional contribution to the England total.
17.4	Number of sites for mineral workings in the East Midlands region in 1988.
18.1	Total permitted area for surface mineral workings in the East Anglia region, indicating the percentage regional contribution to the England total.
18.2	Number of sites for mineral workings in the East Anglia region in 1988.
19.1	Areas of Outstanding Natural Beauty in the South East region, indicating relevant mineral planning authorities and those with mineral workings.
19.2	Total area permitted for surface mineral workings in the South East region, indicating the percentage regional contribution to the England total.
19.3	Total area permitted for the surface disposal of mineral working deposits in the South East region, indicating the percentage regional contribution to the England total.
19.4	Total area of underground permissions in the South East region, indicating the percentage regional contribution to the England total.
19.5	Number of sites for mineral workings in the South East region in 1988.
20.1	National Parks and Areas of Outstanding Natural Beauty in the South West region, indicating relevant mineral planning authorities and those with mineral workings.
20.2	Total area permitted for surface mineral workings in the South West region, indicating the percentage regional contribution to the England total.
20.3	Total area permitted for the surface disposal of mineral working deposits in the South West region, indicating the percentage regional contribution to the England total.
20.4	Total area of underground permissions in the South West region, indicating the percentage regional contribution to the England total.

TABLE TITLE

20.5	Number of sites for mineral workings in the South West region in 1988.
21.1	Total area permitted for surface mineral workings in the West Midlands region, indicating the percentage regional contribution to the England total.
21.2	Total area permitted for the surface disposal of mineral working deposits in the West Midlands region, indicating the percentage regional contribution to the England total.
21.3	Total area of underground permissions in the West Midlands region, indicating the percentage regional contribution to the England total.
21.4	Number of sites for mineral workings in the West Midlands region in 1988.
22.1	Total area permitted for surface mineral workings in the North West region, indicating the percentage regional contribution to the England total.
22.2	Total area permitted for the surface disposal of mineral working deposits in the North West region, indicating the percentage regional contribution to the England total.
22.3	Total area of underground permissions in the North West region, indicating the percentage regional contribution to the England total.
22.4	Number of sites for mineral workings in the North West region in 1988.

INDEX TO FIGURES

FIGURE	TITLE
6.1	Standard regions of England and Mineral Planning Authorities.
7.1	Areas of surface mineral workings permitted in England in 1982 and 1988 - main mineral types.
7.2	Nature of provisions for reclamation of surface mineral workings in England in 1982 and 1988.
7.3	Area and nature of provisions for reclamation of surface mineral workings in England in 1982 and 1988 - main mineral types.
7.4	Areas with reclamation conditions in 1982 and 1988 - indicating nature of conditions and stage of working.
7.5	Areas of land with no provisions for reclamation in 1982 and 1988 - indicating active and ceased workings, areas worked and areas included in the Department's Derelict Land Survey.
7.6	Area and nature of provisions for reclamation of surface mineral workings by standard region in England in 1988.
7.7	Surface mineral workings dependant on imported fill in 1982 and 1988 - main mineral types.
7.8	Land damaged by surface mineral workings and included in the 1988 Derelict Land Survey.
8.1	Areas permitted for the surface disposal of mineral working deposits in 1982 and 1988 - main mineral types.
8.2	Nature of provisions for reclamation relating to the surface disposal of mineral working deposits in England in 1982 and 1988.
8.3	Area and nature of provisions for reclamation for the surface disposal of mineral working deposits in England in 1982 and 1988 - main mineral types.
8.4	Areas for the surface disposal of mineral working deposits with reclamation conditions in 1982 and 1988 indicating nature of conditions and areas not yet affected.
8.5	Surface disposal of mineral working deposits - main mineral types with no provisions for reclamation.
8.6	Areas for the surface disposal of mineral working deposits with no provisions for reclamation in 1982 and 1988 indicating active and ceased workings, areas worked and areas included in the Department's Derelict Land Survey.
8.7	Area and nature of provisions for surface disposal of mineral working deposits by standard region in England in 1988.
8.8	Land damaged by the surface disposal of mineral working deposits and included in the Derelict Land Survey.
9.1	Proposed enduses of surface mineral workings and spoil disposal sites covered by aftercare conditions in 1988.
9.2	Percentage of the total permitted area for surface mineral workings, spoil disposal and areas reclaimed since 1982 which include aftercare conditions.
9.3	Areas of surface mineral workings and spoil disposal covered by aftercare conditions - main mineral types.

FIGURE	TITLE
9.4	Proposed enduses for surface mineral workings and spoil disposal sites with aftercare conditions in England in 1982 - main mineral types.
9.5	Proposed enduses of surface mineral workings and spoil disposal areas with aftercare conditions by standard region in England in 1988.
10.1	Area of underground mining (GDO and mineral planning permissions) recorded in England in 1982 and 1988 - main mineral types.
11.1	Area of land reclaimed between 1982 and 1988 indicating satisfactory and unsatisfactory reclamation standards - main mineral types.
11.2	Area of land reclaimed between 1974 and 1982 indicating satisfactory and unsatisfactory reclamation standards - main mineral types.
11.3	Area of land reclaimed before 1974 indicating satisfactory and unsatisfactory reclamation standards - main mineral types.
11.4	Land reclaimed between 1982 and 1988 as a percentage of the current permitted area - main mineral types.
11.5	Afteruses of land reclaimed after mineral workings in England between 1982 and 1988.
11.6	Area of land reclaimed to different afteruses in England between 1982 and 1988 - main mineral types.
11.7	Afteruses of land reclaimed between 1982 and 1988, following mineral working - main mineral types.
11.8	Choice of afteruse of land reclaimed between 1982 and 1988 by standard region.
11.9	Area of land reclaimed by mineral planning permissions and by "other means" - main mineral types.
11.10	Afteruse of land reclaimed by mineral planning permissions and by "other means" in England between 1982 and 1988.
12.1	Number of sites for surface mineral workings in England in 1982 and 1988 - main mineral types.
12.2	Number of sites for the surface disposal of mineral working deposits in England in 1982 and 1988 - main mineral types.
12.3	Number of sites for underground mining in England in 1982 and 1988 - main mineral types.
12.4	Number of surface mineral workings, areas for the disposal of mineral working deposits and underground mines by standard region in England in 1988.
13.1	Number and area of mineral working deposits permitted in England in 1988 indicating whether operations are under specific planning permission or GDO sites - main mineral types.
15.1	Area of surface mineral workings permitted in the Northern region in 1988 - main mineral types.
15.2	Nature of provisions for reclamation for surface mineral workings in the Northern region in 1988.

FIGURE TITLE

15.3	Area and nature of provisions for reclamation of surface mineral workings in the Northern region in 1988 - main mineral types.
15.4	Surface mineral workings dependant on imported fill in the Northern region in 1988.
15.5	Area permitted for the surface disposal of mineral working deposits in the Northern region in 1988 - main mineral types.
15.6	Nature of provisions for reclamation for the surface disposal of mineral working deposits in the Northern region in 1988.
15.7	Area of mineral workings covered by aftercare conditions in the Northern region in 1988 - main mineral types and proposed enduses.
15.8	Chosen afteruse of land reclaimed by mineral planning permissions and by other means in the Northern region between 1982 and 1988.
15.9	Area of land reclaimed by mineral planning permissions and by other means in the Northern region between 1982 and 1988 - main mineral types.
15.10	Land damaged by mineral workings in the Northern region and included in the Derelict Land Survey in 1988.
16.1	Area of surface mineral workings permitted in Yorkshire and Humberside in 1988 - main mineral types.
16.2	Nature of provisions for reclamation for surface mineral workings in Yorkshire and Humberside in 1988.
16.3	Area and nature of provisions for reclamation of surface mineral workings in Yorkshire and Humberside in 1988 - main mineral types.
16.4	Surface mineral workings dependant on imported fill in Yorkshire and Humberside in 1988.
16.5	Area permitted for the surface disposal of mineral working deposits in Yorkshire and Humberside in 1988 - main mineral types.
16.6	Nature of provisions for reclamation for the surface disposal of mineral working deposits in Yorkshire and Humberside in 1988.
16.7	Area of mineral workings covered by aftercare conditions in Yorkshire and Humberside in 1988 - main mineral types and proposed enduses.
16.8	Area of land reclaimed by mineral planning permissions and by other means in Yorkshire and Humberside between 1982 and 1988 - main mineral types.
16.9	Chosen afteruse of land reclaimed by mineral planning permissions and by other means in Yorkshire and Humberside between 1982 and 1988.
16.10	Land damaged by mineral workings in Yorkshire and Humberside and included in the Derelict Land Survey in 1988.
17.1	Area of surface mineral workings permitted in the East Midlands region in 1988 - main mineral types.
17.2	Nature of provisions for reclamation for surface mineral workings in the East Midlands region in 1988.
17.3	Area and nature of provisions for reclamation of surface mineral workings in the East Midlands region in 1988 - main mineral types.

FIGURE	TITLE
17.4	Surface mineral workings dependant on imported fill in the East Midlands region in 1988.
17.5	Area permitted for the surface disposal of mineral working deposits in the East Midlands region in 1988 - main mineral types.
17.6	Nature of provisions for reclamation for the surface disposal of mineral working deposits in the East Midlands region in 1988.
17.7	Area of mineral workings covered by aftercare conditions in the East Midlands region in 1988 - main mineral types and proposed enduses.
17.8	Area of land reclaimed by mineral planning permissions and by other means in the East Midlands region between 1982 and 1988 - main mineral types.
17.9	Chosen afteruse of land reclaimed by mineral planning permissions and by other means in the East Midlands region between 1982 and 1988.
17.10	Land damaged by mineral workings in the East Midlands region and included in the Derelict Land Survey in 1988.
18.1	Area of surface mineral workings permitted in the East Anglia region in 1988 - main mineral types.
18.2	Nature of provisions for reclamation for surface mineral workings in the East Anglia region in 1988.
18.3	Area and nature of provisions for reclamation of surface mineral workings in the East Anglia region in 1988 - main mineral types.
18.4	Surface mineral workings dependant on imported fill in the East Anglia region in 1988 - main mineral types.
18.5	Area of mineral workings covered by aftercare conditions in the East Anglia region in 1988 - main mineral types and proposed enduses.
18.6	Area of land reclaimed by mineral planning permissions and by other means in the East Anglia region between 1982 and 1988 - main mineral types.
18.7	Chosen afteruse of land reclaimed by mineral planning permissions and by other means in the East Anglia region between 1982 and 1988.
18.8	Land damaged by mineral workings in the East Anglia region and included in the Derelict Land Survey in 1988.
19.1	Area of surface mineral workings permitted in the South East region in 1988 - main mineral types.
19.2	Nature of provisions for reclamation for surface mineral workings in the South East region in 1988.
19.3	Area and nature of provisions for reclamation of surface mineral workings in the South East region in 1988 - main mineral types.
19.4	Surface mineral workings dependant on imported fill in the South East region in 1988.
19.5	Area permitted for the surface disposal of mineral working deposits in the South East region in 1988 - main mineral types.
19.6	Nature of provisions for reclamation for the surface disposal of mineral working deposits in the South East region in 1988.

FIGURE TITLE

19.7	Area of mineral workings covered by aftercare conditions in the South East region in 1988 - main mineral types and proposed enduses.
19.8	Area of land reclaimed by mineral planning permissions and by other means in the South East region between 1982 and 1988 - main mineral types.
19.9	Chosen afteruse of land reclaimed by mineral planning permissions and by other means in the South East region between 1982 and 1988.
19.10	Land damaged by mineral workings in the South East region and included in the Derelict Land Survey in 1988.
20.1	Area of surface mineral workings permitted in the South West region in 1988 - main mineral types.
20.2	Nature of provisions for reclamation for surface mineral workings in the South West region in 1988.
20.3	Area and nature of provisions for reclamation of surface mineral workings in the South West region in 1988 - main mineral types.
20.4	Surface mineral workings dependant on imported fill in the South West region in 1988.
20.5	Area permitted for the surface disposal of mineral working deposits in the South West region in 1988 - main mineral types.
20.6	Nature of provisions for reclamation for the surface disposal of mineral working deposits in the South West region in 1988.
20.7	Area of mineral workings covered by aftercare conditions in the South West region in 1988 - main mineral types and proposed enduses.
20.8	Area of land reclaimed by mineral planning permissions and by other means in the South West region between 1982 and 1988 - main mineral types.
20.9	Chosen afteruse of land reclaimed by mineral planning permissions and by other means in the South West region between 1982 and 1988.
20.10	Land damaged by mineral workings in the South West region and included in the Derelict Land Survey in 1988.
21.1	Area of surface mineral workings permitted in the West Midlands region in 1988 - main mineral types.
21.2	Nature of provisions for reclamation for surface mineral workings in the West Midlands region in 1988.
21.3	Area and nature of provisions for reclamation of surface mineral workings in the West Midlands region in 1988 - main mineral types.
21.4	Surface mineral workings dependant on imported fill in the West Midlands region in 1988.
21.5	Area permitted for the surface disposal of mineral working deposits in the West Midlands region in 1988 - main mineral types.
21.6	Nature of provisions for reclamation for the surface disposal of mineral working deposits in the West Midlands region in 1988.

FIGURE	TITLE
21.7	Area of mineral workings covered by aftercare conditions in the West Midlands region in 1988 - main mineral types and proposed enduses.
21.8	Area of land reclaimed by mineral planning permissions and by other means in the West Midlands region between 1982 and 1988 - main mineral types.
21.9	Chosen afteruse of land reclaimed by mineral planning permissions and by other means in the West Midlands region between 1982 and 1988.
21.10	Land damaged by mineral workings in the West Midlands region and included in the Derelict Land Survey in 1988.
22.1	Area of surface mineral workings permitted in the North West region in 1988 - main mineral types.
22.2	Nature of provisions for reclamation for surface mineral workings in the North West region in 1988.
22.3	Area and nature of provisions for reclamation of surface mineral workings in the North West region in 1988 - main mineral types.
22.4	Surface mineral workings dependant on imported fill in the North West region in 1988.
22.5	Area permitted for the surface disposal of mineral working deposits in the North West region in 1988 - main mineral types.
22.6	Nature of provisions for reclamation for the surface disposal of mineral working deposits in the West Midlands region in 1988.
22.7	Area of mineral workings covered by aftercare conditions in the North West region in 1988 - main mineral types and proposed enduses.
22.8	Area of land reclaimed by mineral planning permissions and by other means in the North West region between 1982 and 1988 - main mineral types.
22.9	Chosen afteruse of land reclaimed by mineral planning permissions and by other means in the North West region between 1982 and 1988.
22.10	Land damaged by mineral workings in the North West region and included in the Derelict Land Survey in 1988.

EXECUTIVE SUMMARY

Introduction

1. The 1988 Survey of Land for Mineral Workings in England describes the nature and extent of mineral workings and spoil tips in England at 1 April 1988. It also records the area of mineral workings reclaimed in England between 1982 and 1988. The information collated was similar to that in the last survey of 1982, but there were some additional tables relating to changes introduced by the Town and Country Planning (Minerals) Act 1981.

2. In total some 114000 hectares of land were affected by permissions for surface mineral extraction or related disposal of mineral wastes. This total represents a decrease of 9000 ha or 8% since 1982. There was considerable variation in the extent to which this land had satisfactory conditions attached to the planning permissions, which would ensure reclamation to an appropriate after-use when mineral activities ceased. Some types of surface mineral workings with a relatively rapid "turnover" of land and where progressive reclamation is possible (eg sand and gravel or opencast coal) had a high proportion of the total permitted area covered by satisfactory conditions. Almost all permissions for opencast coal (99%) and nearly 80% of the area for sand and gravel workings had satisfactory conditions for reclamation. In contrast less than 40% of limestone quarries, 50% of areas for disposal of colliery spoil and only 13% of areas for the disposal of china clay wastes had similar conditions.

3. About 20600 ha of mineral workings and spoil tips had been reclaimed to another use in the 6 year period since 1982. Almost 60% of this land had been reclaimed to agricultural use, and nearly 30% for "amenity" purposes.

4. The survey results mainly present the information within relevant mineral planning authorities, but they also record the extent to which mineral activities take place in National Parks and Areas of Outstanding Natural Beauty.

Surface mineral workings

5. In 1988 the total area permitted for surface mineral workings was 96130 ha, a reduction of 8% since 1982. Of this 52640 ha (55%) had been worked but not yet reclaimed (Paragraph 7.2).

6. Permissions for sand and gravel extraction (29040 ha) were the most extensive, accounting for over 30% of the total area permitted for all minerals. Permissions for the extraction of ironstone, limestone, clay/shale and opencast coal accounted for most of the remainder (Paragraph 7.3 and Fig 7.1).

7. Just under two thirds (63%) of the area permitted for surface workings had satisfactory reclamation conditions (60650 ha). A further 19% (18070 ha) had permissions which included reclamation conditions which were considered to be unsatisfactory, and 17% (15850 ha) had no provisions for reclamation. Less than 2% (1560 ha) had been damaged by mineral workings but was unlikely to be reclaimed because provisions for reclamation were unfulfilled (Paragraph 7.8 and Fig 7.2).

8. The area of surface mineral workings recorded in both this and the 1988 Derelict Land Survey was 1840 ha, a decrease of 960 ha since the position in 1982 (Paragraph 7.23).

9. There was considerable variation between regions in the areas of permissions for surface workings. More than a quarter of all permissions (26000 ha) were within the East Midlands region, while almost 20% were in the South East (18500

ha). Large areas of permissions were also recorded in Yorkshire and Humberside (12200 ha), the South West (11600 ha) and the Northern region (11280 ha). The North West and East Anglia regions in contrast had only 5% (4410 ha) and 6% (5460 ha) of the England total respectively (Paragraph 7.18 and Figure 7.6).

Surface disposal of mineral working deposits

10. The area affected by or authorised for the surface disposal of mineral working deposits (ie spoil tips, silt lagoons etc) was 17970 ha, a decrease of 2% since 1982 (Paragraph 8.2).

11. Most of the area permitted (86%) was associated with spoil from two mineral types; deep mined coal (11120 ha or 62%) and china clay (4330 ha or 24%) (Paragraph 8.3 and Fig 8.1).

12. Only 41% of spoil tips (7330 ha) had permissions which included satisfactory reclamation conditions. Over half (51% or 9330 ha) had no provisions for reclamation (Paragraph 8.5 and Fig 8.2).

13. There were 2460 ha of land which had been affected by spoil tips which was also recorded in the 1988 Derelict Land Survey. This was 14% of the total permitted area for spoil tips (Paragraph 8.16).

14. Over 80% of all permissions for spoil tips were within three regions, the South West (5260 ha or 29%), the East Midlands (4960 or 28%) and Yorkshire and Humberside (4190 ha or 23%) (Paragraph 8.14 and Figure 8.7).

Permissions which include aftercare conditions

15. The power for mineral planning authorities to impose aftercare conditions on new permissions has been available since 22nd February 1982. The survey recorded a total of 881 permissions with aftercare conditions covering an area of 21130 ha. Most of the permissions (93% or 819) were for surface mineral workings. The remaining 7% (62) were for spoil tips (Paragraph 9.3 and Table 9.1).

16. Almost half of all permissions with aftercare conditions were for sand and gravel extraction (46% or 408 permissions), a further 15% were associated with opencast coal (129 permissions) (Paragraph 9.8).

17. Three quarters of the area of surface workings with aftercare conditions had agriculture as the intended land use after reclamation. Amenity uses accounted for 18% and forestry 7%. A proportionally larger percentage of spoil tips had forestry as the final afteruse (16%). Agriculture was still the most common (70%) with amenity uses third (14%) (Paragraph 9.13 and Fig 9.1).

Permissions for underground mining

18. Permissions for underground mining covered over 780000 ha. Almost half (49%) of this area was for vein mineral extraction (379710 ha) (Paragraph 10.5).

19. Some mpas found difficulties in estimating the full extent of permissions for coal mining permitted under the General Development Order, and therefore the figures recorded in the survey were incomplete. Even so, an area of almost 316000 ha of GDO coal mining was recorded, 40% of all underground mines. Coal worked under mineral planning permissions covered 46380 ha (6% of the total) (Paragraph 10.5).

Reclaimed land

20. There were 20590 ha of land reclaimed between 1982 and 1988. Of this 16900 ha (81%) was reclaimed under conditions attached to mineral planning

permissions and 3690 ha by Derelict Land Grant or because of subsequent permissions for other uses of land (Paragraph 11.4).

21. Reclamation of sand and gravel sites was the most common, accounting for 45% of all reclamation (9370 ha). Opencast coal reclamation accounted for 20% of the total (4080 ha) (Paragraph 11.5 and Fig 11.1).

22. Almost 60% of reclaimed land (12090 ha) had an agricultural after-use. Amenity reclamation accounted for 28% (5720 ha), "other uses" 10% and forestry 4% (Paragraph 11.8 and Fig 11.5).

23. Almost 95% of the area reclaimed was considered to be to a satisfactory standard. This was a marginal improvement on the record of reclamation before 1974 and between 1974 and 1982 (Paragraph 11.19).

Number of sites
24. There were a total of 5131 sites associated with mineral workings in England, a reduction of 1734 since 1982 (Paragraph 12.1 and Table 12.1).

25. The number of surface mineral workings decreased by 24% since 1982 to 4343 sites. The most numerous were sand and gravel workings (1533 sites) accounting for 35% of the total; 12% were limestone quarries (521 sites) and 11% were clay/shale sites (496 sites) (Paragraph 12.3 and Table 12.2).

26. There were 474 sites permitted for the surface disposal of mineral working deposits, a reduction of 292 sites (38%) since 1982 (Paragraph 12.7 and Table 12.3). Over half, (54%), were associated with spoil from deep mined coal (Paragraph 12.8).

27. There were 314 underground mines recorded, a reduction of 14% since 1982 (Paragraph 12.12). Two thirds (208 sites) were coal mines (Table 12.4)

Removal of material from mineral working deposits
28. A total of 83 sites were recorded with permissions to rework mineral working deposits, covering a total area of 1270 ha. Of these, 55 sites (1140 ha) were being worked under mineral planning permissions, 28 sites (230 ha) were still being worked under the GDO (Paragraph 13.2 and Table 13.1).

29. The reworking of colliery spoil accounted for over half of the sites (42 sites). Most (33), were worked under mineral planning permissions; 9 were worked under the GDO (Paragraph 13.2)

Reviews under the Town and Country Planning Act 1971
30. At the time of the survey, only 15 mineral planning authorities had begun their reviews. A total of 818 sites or unimplemented permissions had been reviewed of which 175 (21%) had been identified for further action. Only 9 sites had had specific forms of action identified and in all cases this involved voluntary agreements. No formal orders were reported as having been made (Table 14.1).

1. INTRODUCTION

1.1 The 1988 Survey of land for mineral working in England is an update of surveys carried out for the Department of the Environment in 1974 and 1982 by mineral planning authorities. A similar survey is being carried out by mineral planning authorities in Wales on behalf of the Welsh Office.

2. PURPOSE AND SCOPE OF THE SURVEY

2.1 As with the two previous minerals surveys, the 1988 survey is a collation of data on the nature and extent of mineral workings, spoil disposal and reclamation of mineral workings in England. It is also intended to assist in monitoring some of the effects of the Town and Country Planning (Minerals) Act 1981 (the 1981 Act), and to inform future national policies regarding mineral workings and land reclamation. By recording information at local, regional and national level, the Department hopes that the Survey will also assist local authorities as mineral planning authorities to operate more effectively in development control relating to mineral workings. It should be noted that the survey is concerned with the *areas of land affected by mineral activities,* and should not be taken as indicating an assessment of the permitted reserves of workable minerals in different parts of the country. Such assessments are part of the remit of, for example, the Regional Aggregate Working Parties. This survey should be seen as complementary to such work.

2.2 There have been a number of significant changes in legislation and government policy relating to the planning control over mineral workings since the publication of the 1982 survey. In particular, whilst some parts of the 1981 Act came into effect in February 1982, it was fully implemented in May 1986. (Detailed advice on the new provisions introduced by the 1981 Act is provided in the Department's series of Mineral Planning Guidance Notes - MPG's). The 1988 survey therefore asked for additional information on mineral workings subject to aftercare conditions; numbers and areas of mineral working deposits which were being reworked; and progress on the statutory reviews of mineral working sites and any subsequent use of formal orders. The scope of the survey also increased to include oil/gas exploration/appraisal and oil/gas production as two separate categories. Earlier surveys included these two categories under "other minerals".

2.3 There have also been some changes in the responsibilities of local authorities in relation to mineral workings since the 1982 survey. Outside London and the areas of the former metropolitan counties, the County Councils remain the mineral planning authorities. In Greater London and the former metropolitan county council areas, the London borough councils and the metropolitan district councils became responsible for mineral planning matters as from 1 April 1986. The Peak Park Joint Planning Board and the Lake District Special Planning Board remain mpas in their own right. There are now 111 mpas compared with 48 in 1982. Most of the 33 London boroughs, however, do not have mineral workings within their areas.

3. ORGANISATION OF THE SURVEY

3.1 The Department sent the Chief Planning Officer of each mineral planning authority a survey form and guidance notes in January 1988. These set out the aims and objectives of the survey, with guidance on how to complete the forms. To assist the local authorities in completing the forms and to provide a basis for comparison with earlier surveys, the general format of the forms and the nature of the required response was similar to the 1982 survey wherever possible. Information relating to changes brought about by the 1981 Minerals Act was included as additional tables. A copy of the survey form, guidance notes and a glossary of terms is at Appendix 1.

3.2 The base date of the survey was 1 April 1988. The mineral planning authorities were asked to return the completed forms to the Minerals Division, Department of the Environment by 1 October 1988.

3.3 The survey includes all mineral workings covered by planning permissions, those worked under the GDO and any unauthorised workings. Sites restored before 1st April 1982 and those considered to have naturally revegetated were excluded.

3.4 Areas of land which as a result of the mineral working industry had become derelict were included both in this and the Department's concurrent 1988 Derelict Land Survey.

4. THE RESPONSE

4.1 All mineral planning authorities, with the exception of North Yorkshire and Oxfordshire, provided returns. The amount of time which mpa's had to spend on the survey and their views on the accuracy of the results varied. This in part reflected the differences in the nature and extent of each authority's own database of mineral planning information. The general comment made by most mineral officers was that the information provided for the 1988 survey was more accurate than in 1982.

4.2 To enable national and regional comparisons to be made with the 1982 survey, relevant data provided by North Yorkshire and Oxfordshire in 1982 has been included in this survey. On the basis of the 1988 returns from all other mpa's it is considered that this approach does not unacceptably distort the information. However, there is no data from these two authorities in this survey on the extent of reworking of mineral deposits, aftercare conditions or the stage reached in the review of mineral workings. The area of land reclaimed within the two authorities between 1982 and 1988 is not included in this survey.

4.3 The Department is most grateful to the staff of all the mpa's who provided returns for the 1988 survey.

5. COMPARISON WITH 1982 AND 1974 SURVEYS

5.1 The scope of the 1988 survey and hence the number of tables have expanded to monitor the effects of the 1981 Minerals Act. There was a more detailed breakdown of reclaimed land to identify the enduse, and separation of areas reclaimed under the planning permission granted for the working of the mineral and land reclaimed by some other mechanism such as Derelict Land Grant or planning permissions granted for subsequent developments.

5.2 Comparisons are made between the extent and nature of mineral workings in 1982 and 1988. Where appropriate, reference has also been made to the 1974 survey. However because of differences in the breakdown of data in the 1974 survey, these comparisons are more restricted in number.

6. PRESENTATION OF RESULTS

6.1 The results of the 1988 survey have been presented in three volumes. This volume, Volume 1 provides a commentary on the survey results, and volumes 2 and 3 provide the detailed figures.

6.2 **Volume 1** contains an analysis of the national and regional distribution of mineral workings in England and compares the results with earlier surveys. The map, Figure 6.1, shows the standard regions used in the analyses and their relation to individual mpa's. Sections 7 to 14 of the following text present the analysis at national level, whilst Sections 15 to 22 are regional commentaries.

6.3 **Volume 2** provides the detailed 1988 figures for individual mineral planning authorities, together with national and regional summaries. This volume also provides tables of mineral workings in individual National Parks and AONB's.

6.4 **Volume 3** presents the same data as volume 2 but expressed by mineral type. Hence for each mineral type the areas affected in each mineral planning authority within the eight standard regions of England are presented, together with national and regional totals.

FIGURE 6.1

STANDARD REGIONS OF ENGLAND AND MINERAL PLANNING AUTHORITIES

PART 1　NATIONAL COMMENTARY

7. SURFACE MINERAL WORKINGS

Introduction

[Information on surface mineral workings was taken from Tables 1A, 1B and 1C of the survey form and has been collated into Tables A1 to A8 in Volumes 2 and 3 of the survey results].

7.1 All county councils, all metropolitan district councils with the exception of Liverpool and Birmingham and seven of the thirty-three London boroughs had surface mineral workings within their area.

7.2 Over 96100 ha of land were recorded as having permissions for surface mineral workings in England in 1988. This marked a reduction in area of 8% (8670 ha) since 1982. Of this, 52640 ha (55%) had been worked but not reclaimed and the remainder was unworked permissions. This compares with 53400 ha worked but not reclaimed in 1982.

7.3 Sand and gravel remained the mineral type with the largest area of permissions for surface mineral workings, covering over 29000 ha (Table 7.1). This was over 30% of the total permitted area for all surface working (Figure 7.1). Ironstone permissions covered 14420 ha, limestone 11490 ha, clay/shale 10090 ha and opencast coal 8420 ha. These five minerals accounted for three quarters of the permitted area for all minerals. However, ironstone extraction in England ceased in 1979 and over 13000 ha of the area recorded remains unworked. Most was considered unlikely to be worked in the future for the original mineral although the permissions remain valid.

TABLE 7.1 AREA OF LAND PERMITTED FOR SURFACE MINERAL WORKINGS IN ENGLAND IN 1982 AND 1988.

Hectares

MINERAL TYPE	AREA IN 1982	AREA IN 1988	CHANGE IN AREA
CHALK	3350	3660	+310
CHINA CLAY	3180	2380	-800
CLAY/SHALE	12820	10090	-2730
OPENCAST COAL	7760	8420	+660
GYPSUM/ANHYDRITE	650	810	+160
IGNEOUS ROCK	2640	2240	-400
IRONSTONE	15700	14420	-1280
LIMESTONE/DOLOMITE	11880	11490	-390
OIL/GAS EXPLORATION	–	80	+80
OIL/GAS PRODUCTION	–	60	+60
SAND & GRAVEL	31960	29040	-2920
INDUSTRIAL SAND	2150	2130	-20
SANDSTONE	3420	2940	-480
SLATE	330	310	-20
VEIN MINERALS	1150	1540	+390
OTHER MINERALS	7820	6530	-1290
TOTAL	104800	96130	-8670

Notes: Areas are rounded to the nearest 10 hectares.
Oil/gas exploration and production were included with "other minerals" in 1982.

7.4 There was a small decrease in the permitted area for four out of the five most widespread minerals, opencast coal being an exception (Table 7.1). As a result the relative proportions of each mineral changed very little since 1982, as illustrated in Figure 7.1.

7.5 Reclamation conditions were attached to 82% of the total permitted area in the 1988 survey. Over 16% had no provisions for reclamation and less than 2% had been damaged by mineral working and was unlikely to be reclaimed. Comparison is made with 1982 data in Figure 7.2, and a breakdown of the main mineral types in Figure 7.3. This is discussed in further detail below.

Permissions which include reclamation conditions

[Information taken from Table 1C of the survey form]
7.6 Over 78720 ha of surface mineral workings had permissions which included reclamation conditions, 82% of all permissions. This was a smaller area than in 1982 (85540 ha), but accounted for the same percentage of the total area (Figure 7.2).

7.7 The main change since 1982 has been a reduction in the areas of unworked permissions with satisfactory reclamation conditions (Figure 7.4). In 1988, 39580 ha (50%) of the area with reclamation conditions had not yet been worked. This compares with 44180 ha (52%) in 1982. It should be noted however that the data for 1982 does not include china clay permissions (an additional 1090 ha) as worked and unworked areas were not separately categorised.

7.8 Mineral planning authorities were asked to determine whether the reclamation conditions were adequate to ensure satisfactory reclamation, taking account of the enforceability of conditions, current restoration progress and the record or attitude of the operator. Mpa's considered that 63% of the total area of surface mineral permissions had satisfactory reclamation conditions (60650 ha) which represented approximately 77% of the area with any type of reclamation conditions. Permissions with unsatisfactory reclamation conditions covered 18070 ha, 19% of the total permitted area. The proportion of permissions with satisfactory reclamation conditions was slightly lower than the 66% recorded in 1982 (Figure 7.2). This small decline may reflect a change in perception by the local authorities as to what constitutes satisfactory conditions. Comments made by a number of mpa's indicated that some sites recorded in 1982 as having satisfactory reclamation conditions were reclassified as unsatisfactory in this survey.

7.9 Minerals worked over long timescales, and which are thought therefore to have proportionally more old permissions, had in general a higher proportion of unsatisfactory or no reclamation conditions (Figure 7.3). Hardrock (in particular limestone and sandstone), and chalk quarries had a low percentage of the total permitted area covered by satisfactory conditions (33% - 38%). These figures may also reflect the technical difficulty of devising satisfactory reclamation conditions for large, deep and long life quarries. Only 32% of the total area of china clay workings, 20% of vein minerals and 7% of slate had satisfactory reclamation conditions (Table 7.2).

7.10 This contrasts markedly with opencast coal workings, for which 99% of the total permitted area had satisfactory conditions. The same result for opencast coal was obtained in 1982. High percentages were also recorded for permissions for oil/gas exploration (96%) and production (97%), industrial sand (82%) and sand and gravel (78%).

FIGURE 7.1 AREAS OF SURFACE MINERAL WORKINGS PERMITTED IN ENGLAND IN 1982 AND 1988 – MAIN MINERAL TYPES

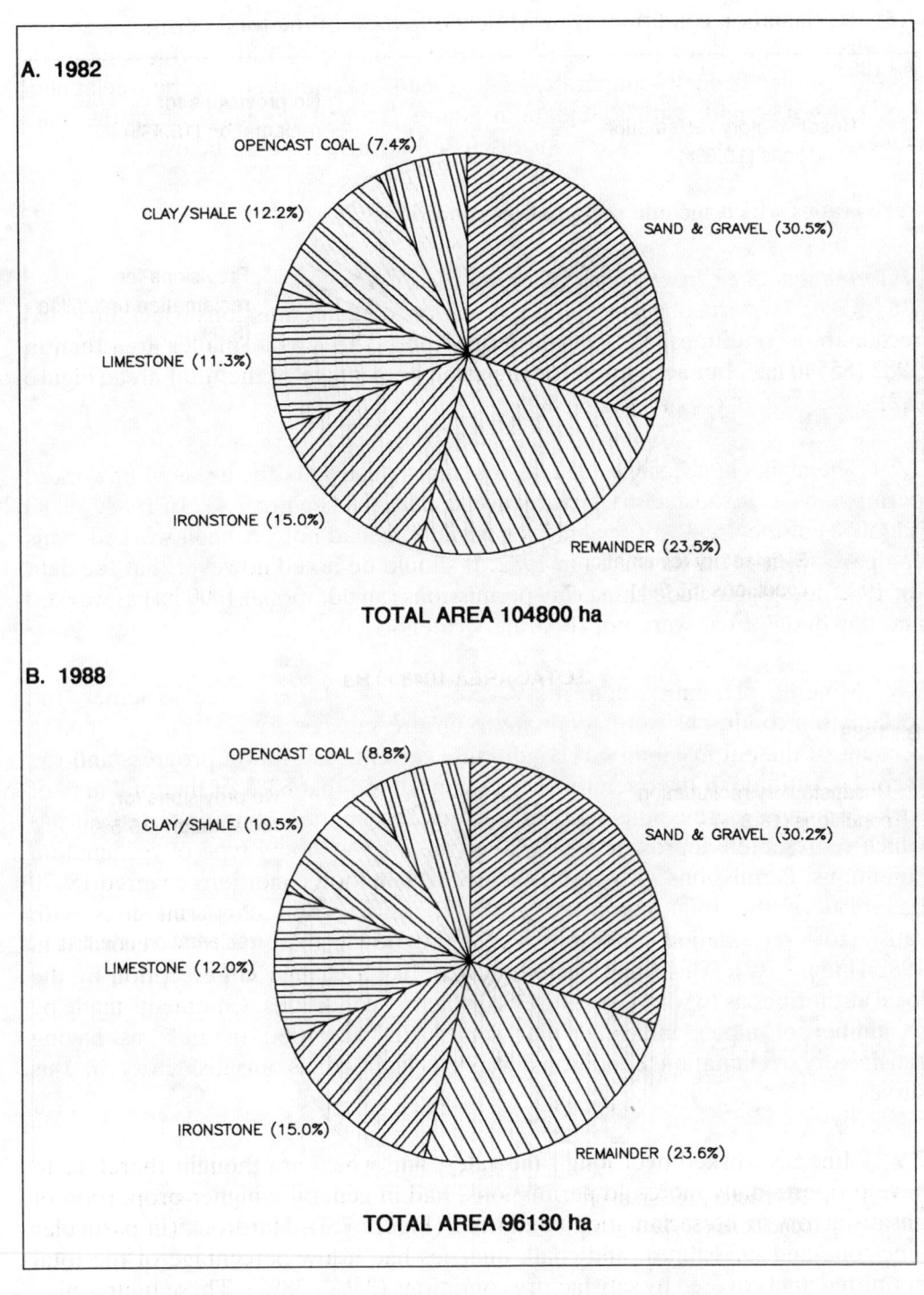

FIGURE 7.2 NATURE OF PROVISIONS FOR RECLAMATION OF SURFACE MINERAL WORKINGS IN ENGLAND IN 1982 AND 1988

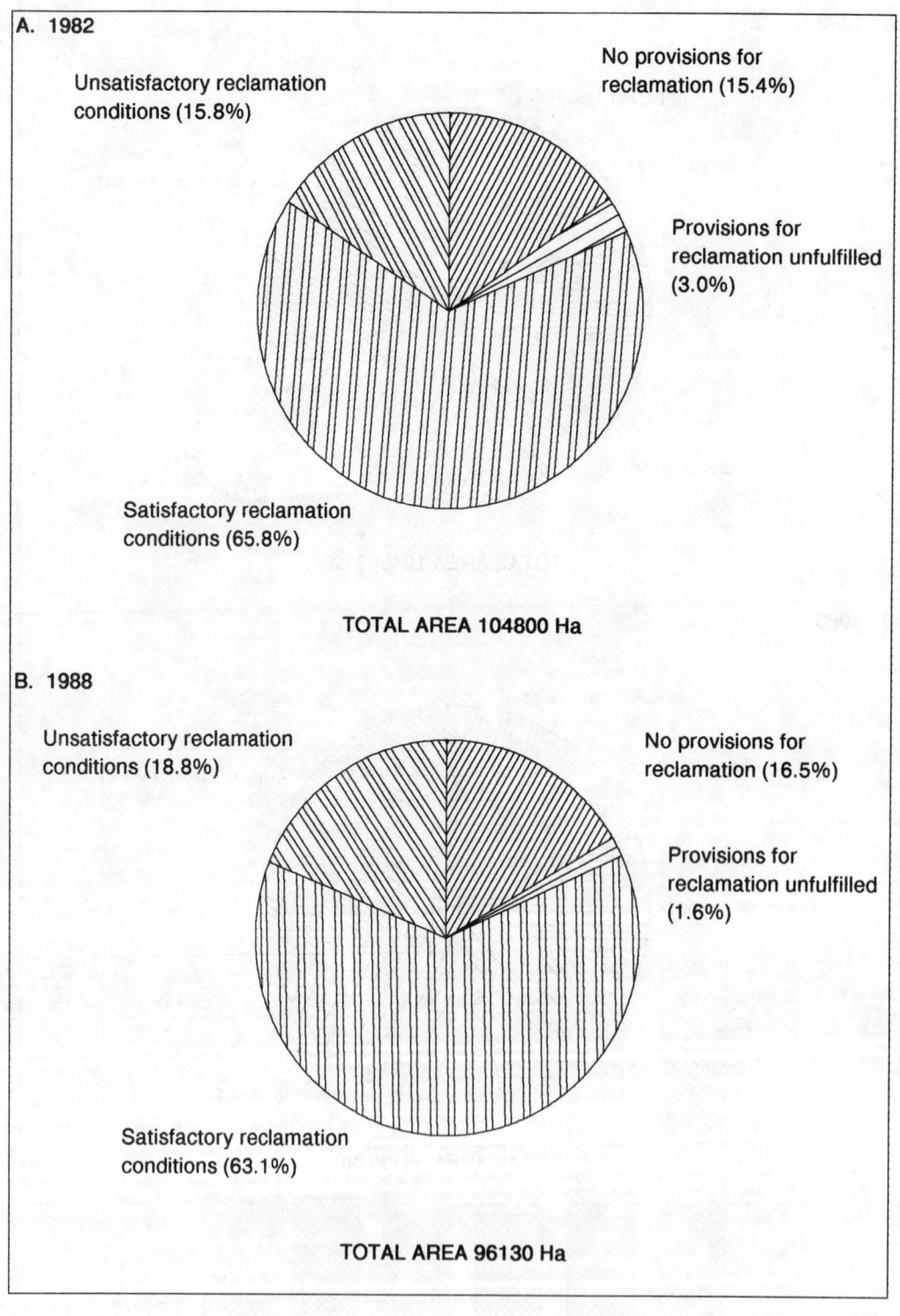

FIGURE 7.3 AREA AND NATURE OF PROVISIONS FOR RECLAMATION OF SURFACE MINERAL WORKINGS IN ENGLAND IN 1982 AND 1988 – MAIN MINERAL TYPES.

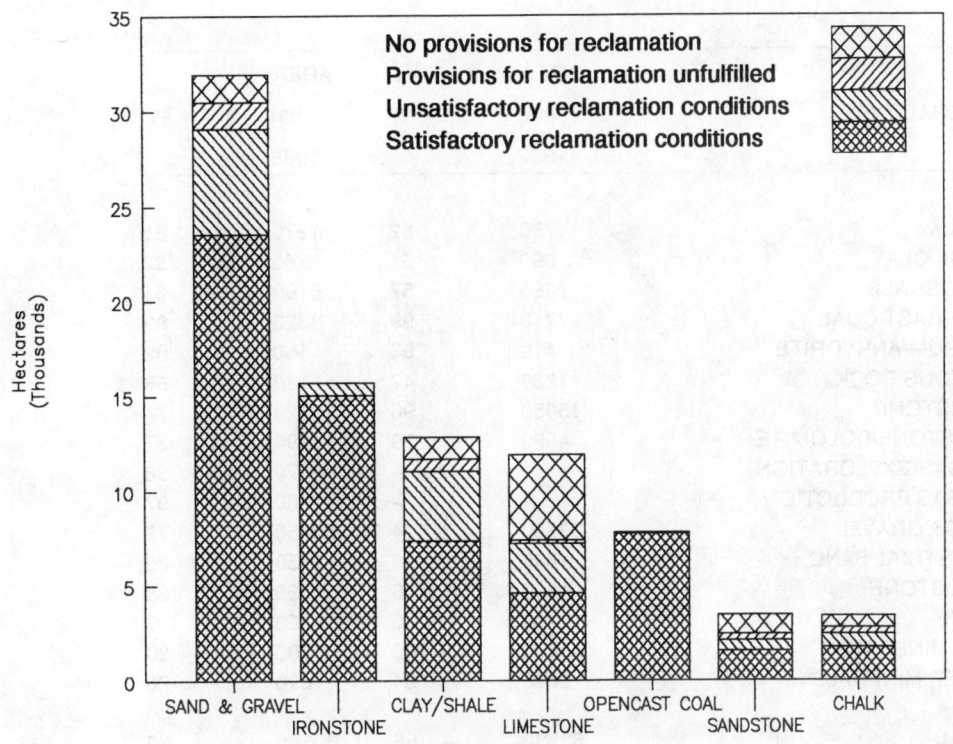

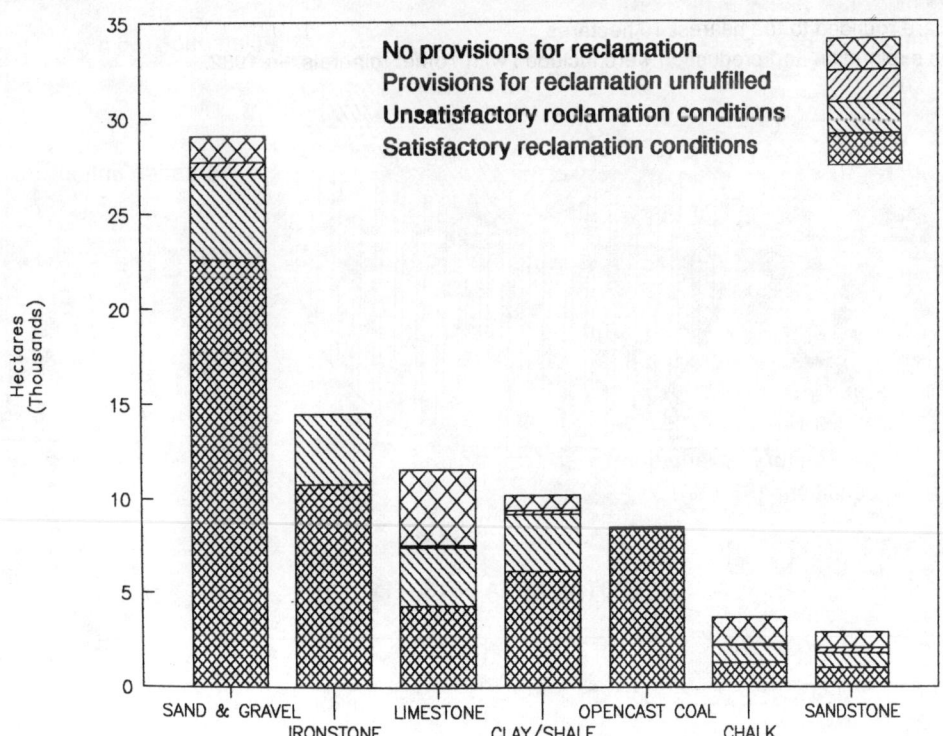

TABLE 7.2 AREA OF SURFACE WORKINGS WITH SATISFACTORY RECLAMATION CONDITIONS IN 1982 AND 1988 INDICATING THE PERCENTAGE OF THE TOTAL PERMITTED AREA

MINERAL TYPE	AREA (ha) 1982	%	AREA (ha) 1988	%
CHALK	1730	52	1270	35
CHINA CLAY	990	31	760	32
CLAY/SHALE	7350	57	6100	61
OPENCAST COAL	7710	99	8320	99
GYPSUM/ANHYDRITE	410	63	560	68
IGNEOUS ROCK	1230	47	1180	53
IRONSTONE	15050	96	10720	74
LIMESTONE/DOLOMITE	4580	39	4290	37
OIL GAS EXPLORATION	-	-	70	96
OIL/GAS PRODUCTION	-	-	60	97
SAND & GRAVEL	23530	74	22550	78
INDUSTRIAL SAND	1640	76	1750	82
SANDSTONE	1560	46	1020	33
SLATE	0	0	20	7
VEIN MINERALS	470	40	300	20
OTHER MINERALS	2690	34	1690	26
TOTAL	68930	66	60580	63

Notes:
Areas are rounded to the nearest 10 hectares
Oil/gas exploration and production were included with "other minerals" in 1982.

FIGURE 7.4 AREAS WITH RECLAMATION CONDITIONS IN 1982 AND 1988 – INDICATING NATURE OF CONDITIONS AND STAGE OF WORKING

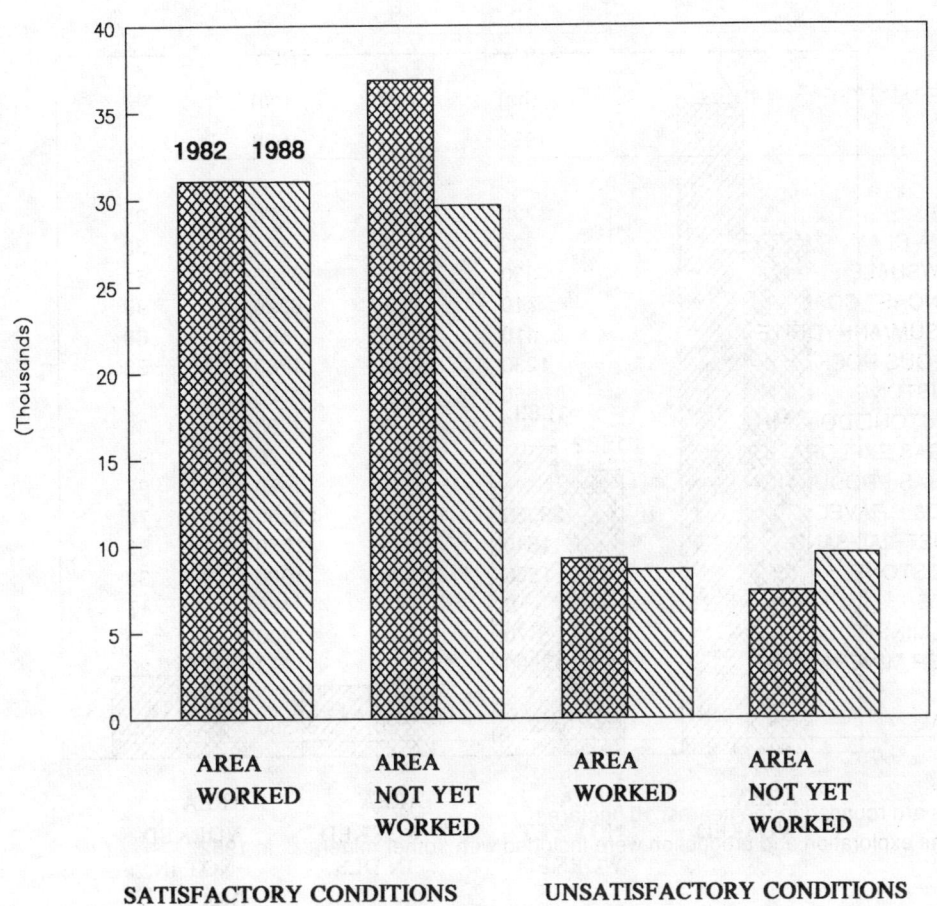

Permissions with no provisions for reclamation

[Figures recorded on Table 1A of the survey Form]

7.11 There were 15850 ha of surface mineral workings with no reclamation conditions nor other provision for reclamation (16%). This compares with 16110 ha (15%) in 1982, a reduction of 260 ha.

7.12 Most of the area with no provisions for reclamation (86%) were still active (Figure 7.5). Over 9000 ha of these active sites had been worked, 4400 ha remained to be worked. The figures for 1982 do not include data for china clay because it was not possible to subdivide areas into worked and unworked areas. Direct comparison with 1988 data therefore gives the impression of a large overall increase in area with no provisions for reclamation.

7.13 Two mineral types accounted for over half of all workings with no provisions for reclamation (see Figure 7.3), "other minerals" (4300 ha) and limestone (4000 ha). The large figure for "other minerals" corresponds largely to peat workings within Doncaster MBC and Somerset CC (in total over 3400 ha). Significant areas of chalk (1420 ha), china clay (1490 ha) and sand and gravel (1370 ha) were also recorded (Table 7.3).

FIGURE 7.5 AREAS OF LAND WITH NO PROVISIONS FOR RECLAMATION IN 1982 AND 1988 – INDICATING ACTIVE AND CEASED WORKINGS, AREAS WORKED AND AREAS INCLUDED IN THE DEPARTMENT'S DERELICT LAND SURVEY.

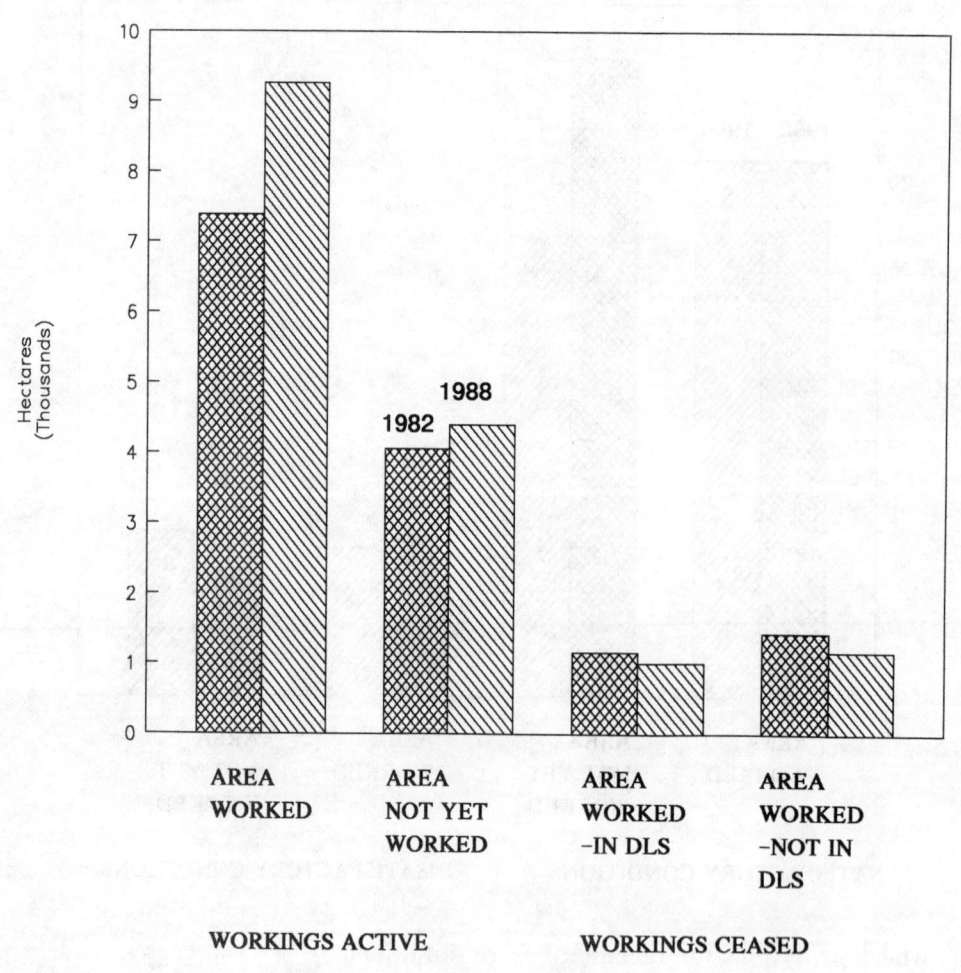

TABLE 7.3 AREA OF SURFACE MINERAL WORKINGS WITH NO PROVISIONS FOR RECLAMATION IN 1982 AND 1988 INDICATING THE PERCENTAGE OF THE TOTAL PERMITTED AREA

Hectares

MINERAL TYPE	AREA 1982	%	AREA 1988	%
CHALK	610	18	1420	39
CHINA CLAY	2090	66	1490	62
CLAY/SHALE	1160	9	770	8
OPENCAST COAL	20	<1	10	<1
GYPSUM/ANHYDRITE	100	15	120	14
IGNEOUS ROCK	800	30	410	18
IRONSTONE	20	<1	20	<1
LIMESTONE/DOLOMITE	4510	38	4000	35
OIL/GAS EXPLORATION	–	–	<10	4
OIL/GAS PRODUCTION	–	–	<10	3
SAND & GRAVEL	1470	5	1370	5
INDUSTRIAL SAND	210	10	140	7
SANDSTONE	1010	30	910	31
SLATE	320	97	260	84
VEIN MINERALS	140	12	650	42
OTHER MINERALS	3700	47	4300	66
TOTAL	16110		15860	

Note: Areas are rounded to the nearest 10 hectares.

Areas where provisions for reclamation are unfulfilled

[Figures recorded in Table 1B of the survey]

7.14 A small proportion of mineral workings which nominally have reclamation conditions attached (or which had some other arrangement for reclamation), cause a problem because the provisions for reclamation are unfulfilled. Examples included sites worked under planning permission which required fill prior to reclamation where filling had subsequently become an unrealistic prospect. Also included were sites where voluntary agreements for reclamation between the mineral operator and the land owner had proved ineffectual.

7.15 Excluded from this category were sites where reclamation had been carried out, but to an unsatisfactory standard (these were included in Tables 4A and 4B). Also excluded were sites where working had ceased within the previous five years and the efficiency of the reclamation conditions had not been adequately tested.

7.16 Only 1.6% of the total area of surface mineral workings were recorded in this category - a total area of 1560 ha. This was less than half the area recorded in 1982 (3140 ha).

7.17 Mineral types most affected were sand and gravel (640 ha), sandstone (280 ha) and clay/shale (200 ha) (see Figure 7.3).

Regional comparisons

7.18 Figure 7.6 illustrates the regional variability in the areas of permissions and the nature of provisions for reclamation for surface mineral workings. The East Midlands region had the largest area of permissions for surface mineral workings (26180 ha), 27% of the England total. Large areas of permissions were also recorded in the South East (18500 ha or 20%), Yorkshire and Humberside (12200 ha or 13%), the South West (11600 ha or 12%) and the Northern region (11280 ha or 12%). In contrast, the North West region contained permissions covering only 4410 ha (less than 5%), the smallest total for a region in England.

7.19 There was also great regional variability in the nature of the provisions for reclamation of mineral workings. Almost 95% of the permitted area in East Anglia and 78% in the Northern region was covered by satisfactory reclamation conditions. In contrast, the South West region recorded only 27% of the permitted area as being covered by satisfactory reclamation conditions. Almost 50% had no provisions for reclamation, whilst a further 23% was covered by unsatisfactory reclamation conditions.

7.20 Yorkshire and Humberside and the North West regions also had only a relatively small percentage of the permitted area covered by satisfactory reclamation conditions. Over 30% of the permitted area in Yorkshire and Humberside had no provisions for reclamation, whilst less than half had satisfactory reclamation conditions. Similarly 25% of permissions in the North West region had no provisions for reclamation and only 52% had satisfactory conditions.

Areas dependant on fill

[Figures recorded in Table 1C of survey]

7.21 Almost 9500 ha of surface mineral workings were recorded as being dependant on imported fill to achieve reclamation, a reduction of 1500 ha since 1982. Fill included both mineral working deposits not generated on site and controlled waste (household, industrial or commercial waste). The reduction in total area requiring imported fill was approximately proportional to the reduction in total permitted area since 1982, with 10.5% of all surface workings requiring fill in 1982, 9.8% in 1988.

7.22 Half (50%) of the total area requiring fill involved sand and gravel workings (4710 ha) (Figure 7.7). This was a smaller proportion than in 1982 when sand and gravel accounted for 58% (6400 ha), of the total area requiring fill. Clay/shale was the second largest area, 2860 ha (30% of total), an increase of almost 4% since 1982.

Areas included in the 1988 Derelict Land Survey

[Figures recorded in tables 1A and 1B of survey]

7.23 Land derelict as a result of post 1947 mineral workings were recorded in both this and the Departments' Derelict Land Survey. A total area of 1840 ha were recorded, 2% of the total permitted area of surface workings. 1000 ha were sites which had no reclamation conditions, and 840 ha were sites where reclamation commitments were unfulfilled (Fig. 7.8). This was a decrease in area of 960 ha since 1982.

7.24 The minerals with the largest derelict areas were sand and gravel (440 ha), chalk (410 ha), limestone (270 ha), clay/shale (250 ha) and igneous rock (80 ha). However in comparison with 1982, all these figures represented a decrease except for chalk. The increase in area of derelict chalk workings was a result of a large increase in the area recorded in Kent.

FIGURE 7.6 AREA AND NATURE OF PROVISIONS FOR RECLAMATION OF SURFACE MINERAL WORKINGS BY STANDARD REGION IN ENGLAND IN 1988.

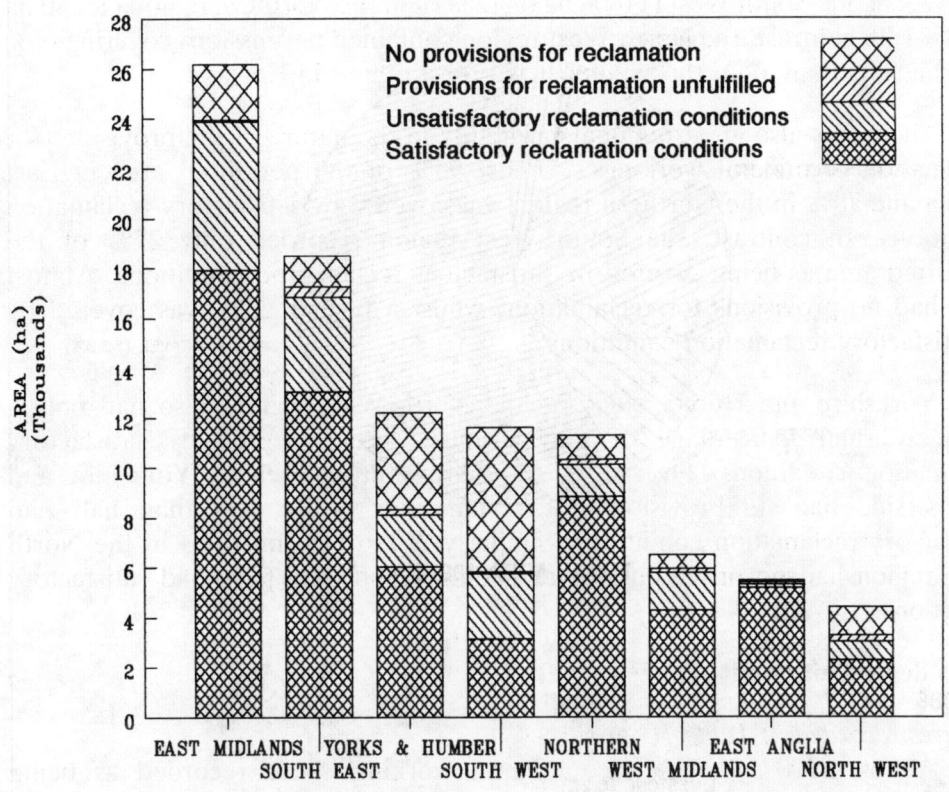

FIGURE 7.7 SURFACE MINERAL WORKINGS DEPENDANT ON IMPORTED FILL IN 1982 AND 1988 – MAIN MINERAL TYPES.

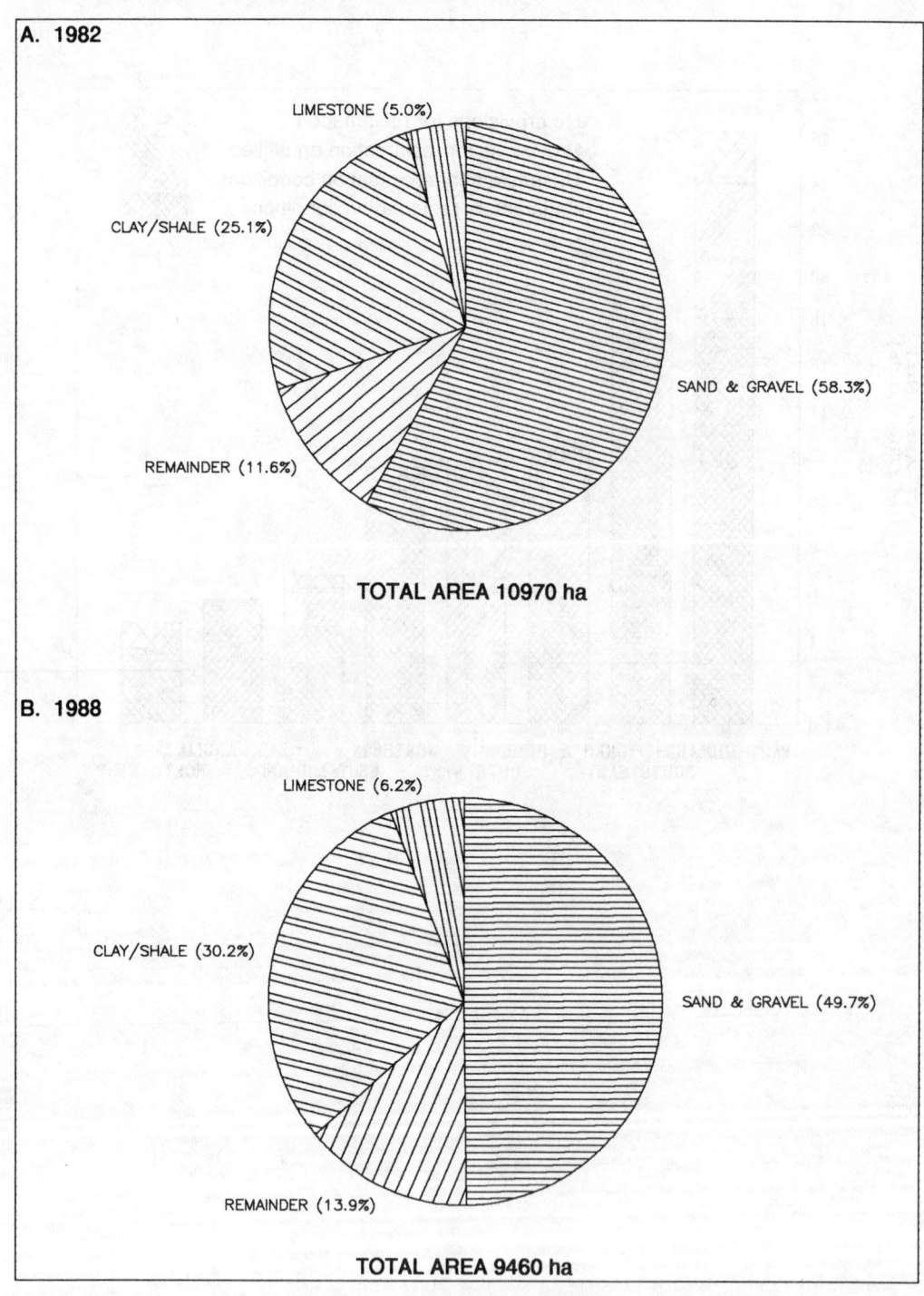

FIGURE 7.8 LAND DAMAGED BY SURFACE MINERAL WORKINGS AND INCLUDED IN THE 1988 DERELICT LAND SURVEY.

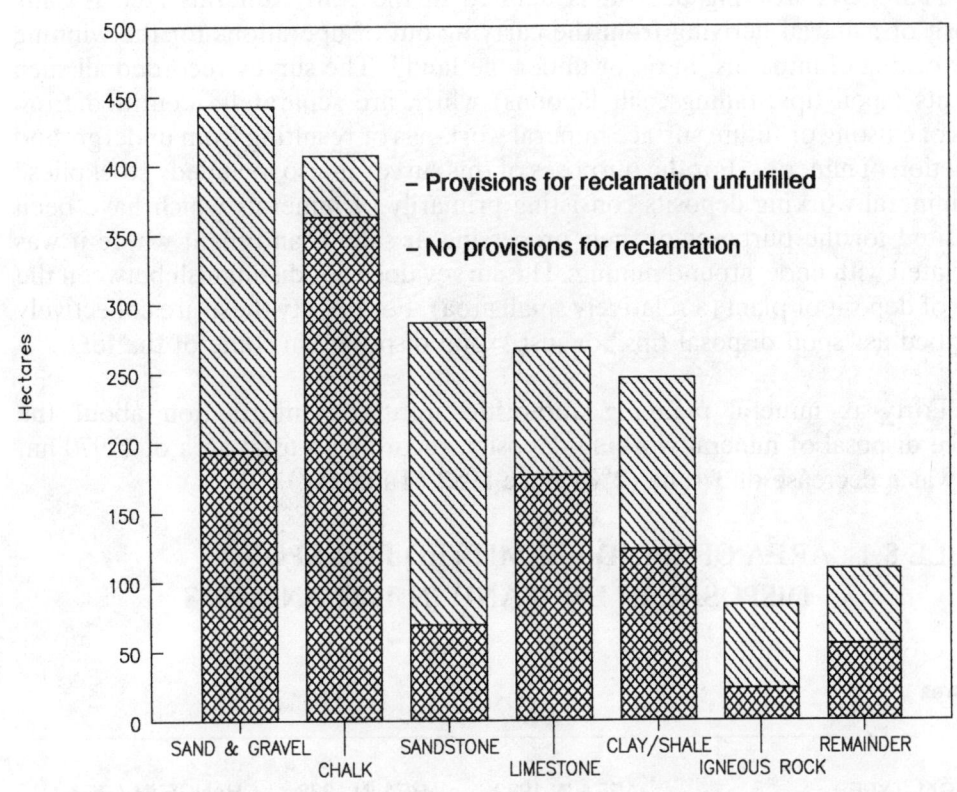

8. SURFACE DISPOSAL OF MINERAL WORKING DEPOSITS

Introduction

[Information about the surface disposal of mineral working deposits was taken from Tables 1A, 1B and 1C of the survey returns and has been collated into Tables B1 to B8 in Volumes 2 and 3 of the survey results].

8.1 A mineral working deposit is defined in the 1981 Minerals Act as "any deposit of material deriving from the carrying out of operations for the winning and working of minerals, in on or under the land". The survey recorded all such deposits (spoil tips, tailings, silt lagoons) which are separately identified from areas of existing or future surface mineral workings or resulting from underground extraction of minerals. For the purposes of this survey it also included "stockpiles" (ie "mineral working deposits consisting primarily of minerals which have been deposited for the purposes of their processing or sale"), and plant where it was associated with underground mining. The survey does not distinguish between the types of deposit or plant (a relatively small area). For brevity these are collectively described as "spoil disposal tips" or just "spoil disposal" in some of the text.

8.2 Forty-six mineral planning authorities returned information about the surface disposal of mineral working deposits, recording a total area of 17970 ha. This was a decrease of 370 ha (2%) since 1982 (Table 8.1).

TABLE 8.1 AREA OF LAND PERMITTED FOR SPOIL DISPOSAL IN ENGLAND IN 1982 AND 1988

Hectares

MINERAL TYPE	AREA IN 1982	AREA IN 1988	CHANGE IN AREA
CHINA CLAY	2810	4330	+1520
CLAY/SHALE	260	350	+90
DEEP MINED COAL	12370	11120	−1250
OPENCAST COAL	30	130	+100
GYPSUM/ANHYDRITE	80	0	−80
IGNEOUS ROCK	550	530	−20
IRONSTONE	100	70	−30
LIMESTONE/DOLOMITE	1020	780	−240
SAND & GRAVEL	120	140	+20
INDUSTRIAL SAND	<10	20	+10
SANDSTONE	220	60	−160
SLATE	170	70	−100
VEIN MINERALS	560	320	−240
OTHER MINERALS	50	40	−10
TOTAL	18340	17970	−370

Notes: Areas are rounded to the nearest 10 hectares

8.3 Spoil derived from the extraction of two mineral types dominated the area recorded within this category; deep mined coal, with permissions covering 11120 ha and china clay (4330 ha). These two minerals accounted for 62% and 24% of the total area respectively. Permissions for spoil disposal from limestone and igneous rock quarries were also extensive covering 780 ha (4%) and 530 ha (3%) respectively (Figure 8.1).

FIGURE 8.1 AREAS PERMITTED FOR THE SURFACE DISPOSAL OF MINERAL WORKING DEPOSITS IN 1982 AND 1988 – MAIN MINERAL TYPES.

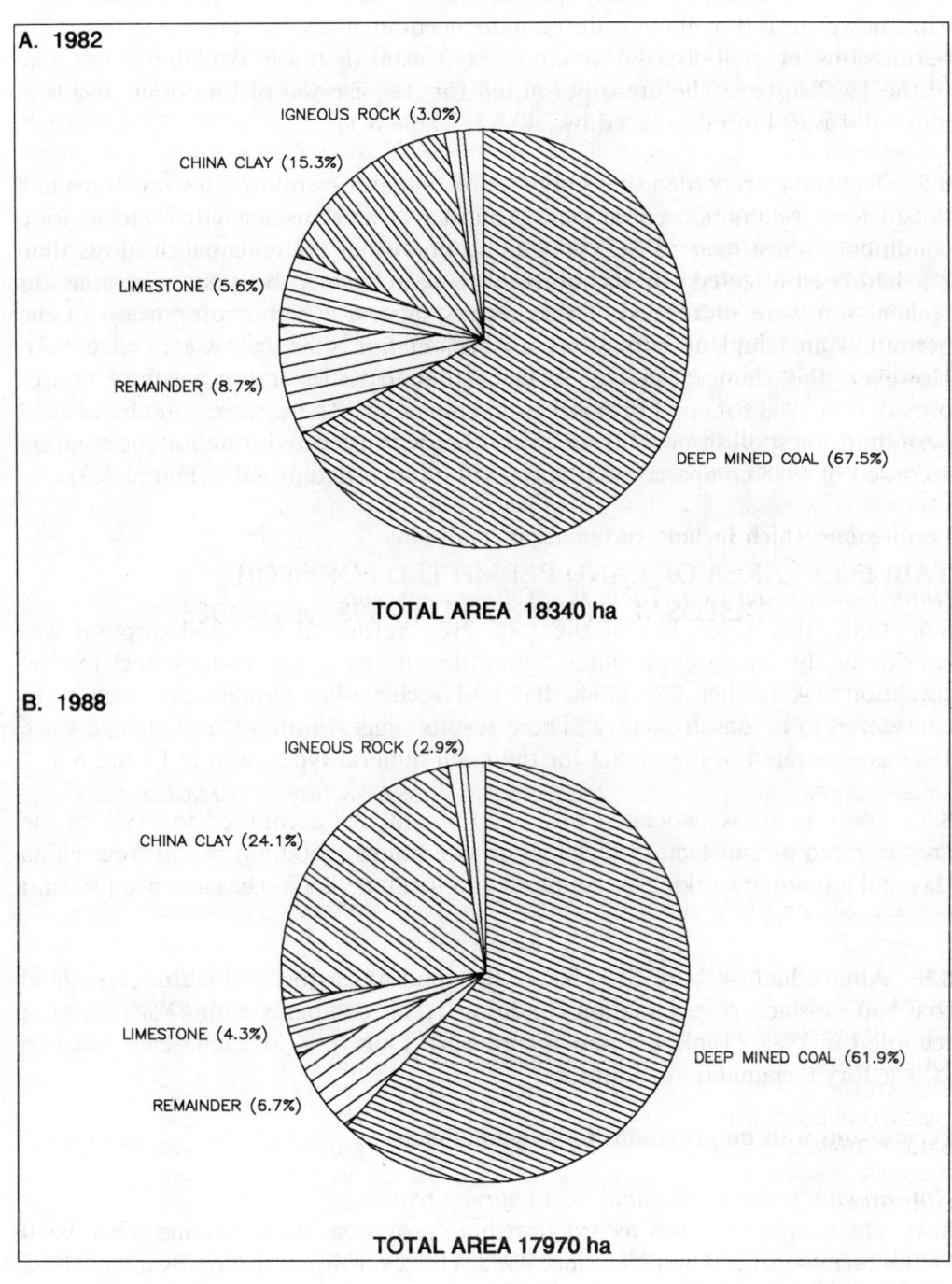

8.4 The total area affected by or authorised for spoil disposal from deep mined coal decreased by 1250 ha (10%) since 1982. This is likely to reflect the closure of many collieries together with the reclamation of significant affected areas, largely under the Derelict Land Grant Programme (2300 ha of colliery spoil reclaimed with Derelict Land Grants since 1982 were recorded in the Derelict Land Survey). The area recorded for china clay spoil tips increased by 1520 ha (54%) since 1982. This, however, is thought to reflect a more accurate assessment of the total area of permissions for spoil disposal for china clay wastes than was available at the time of the 1982 survey. The areas permitted for the disposal of limestone and vein mineral wastes both decreased by 240 ha (Table 8.1).

8.5 The survey recorded that only 41% of the area permitted for spoil tips had satisfactory reclamation conditions. Almost 6% had unsatisfactory reclamation conditions, whilst over half (51%) had no provisions for reclamation. Less than 2% had been affected but was unlikely to be reclaimed because provisions for reclamation were unfulfilled. These results suggest a higher proportion of the permitted area had no provisions for reclamation than in 1982 (Figure 8.2). However, this change can largely be attributed to the more accurate figures provided in 1988 for spoil disposal from china clay. There were 3700 ha of land permitted for spoil disposal from china clay with no provisions for reclamation recorded in 1988 compared with only 1650 ha recorded in 1982 (Figure 8.3).

Permissions which include reclamation conditions

[Information taken from Table 1C of the survey return]
8.6 Only 41% (7330 ha) of the total area permitted for spoil disposal was considered by mineral planning authorities to have satisfactory reclamation conditions. A further 6% (1040 ha) had reclamation conditions which were considered to be unsatisfactory. These results suggest little overall change since 1982 as illustrated in Figure 8.3 for the main mineral types, and in Figure 8.4.

8.7 Spoil tip areas associated with deep mined coal accounted for 75% of the area covered by satisfactory reclamation conditions (5500 ha). Spoil from china clay and limestone workings accounted for a further 8% (560 ha) and 5% (380 ha) respectively (Table 8.2).

8.8 Almost half (49% or 5500 ha) of the spoil tips associated with deep mined coal had satisfactory reclamation conditions. This compares with 43% (5320 ha) recorded in 1982 (Table 8.2 and Figure 8.3). Only 13% of china clay tips had satisfactory reclamation conditions.

Permissions with no provisions for reclamation

[Information taken from Table 1A of survey return]
8.9 The area of spoil tips recorded with no provisions for reclamation was 9330 ha, an increase of 220 ha (2%) since 1982. However direct comparison with 1982 data is misleading because of the revised figure for china clay (discussed in paragraphs 8.4 and 8.5). If china clay is excluded from the comparison there was a reduction of 1890 ha to the total area permitted for spoil tips with no reclamation conditions. The major contributory factor for this decrease was the reduction in the area of deep mined coal permissions (1390 ha). Reductions within this category of 130 ha of igneous rock permissions, 130 ha of sandstone and 90 ha of vein minerals permissions were also recorded (Table 8.3).

8.10 Despite the large reduction in the area of colliery spoil since 1982, there were still 4960 ha recorded as having no provisions for reclamation, accounting for over half of the total area within this category. China clay tips accounted for a further 40% (3700 ha) (Figure 8.5).

FIGURE 8.2 NATURE OF PROVISIONS FOR RECLAMATION RELATING TO THE SURFACE DISPOSAL OF MINERAL WORKING DEPOSITS IN ENGLAND IN 1982 AND 1988.

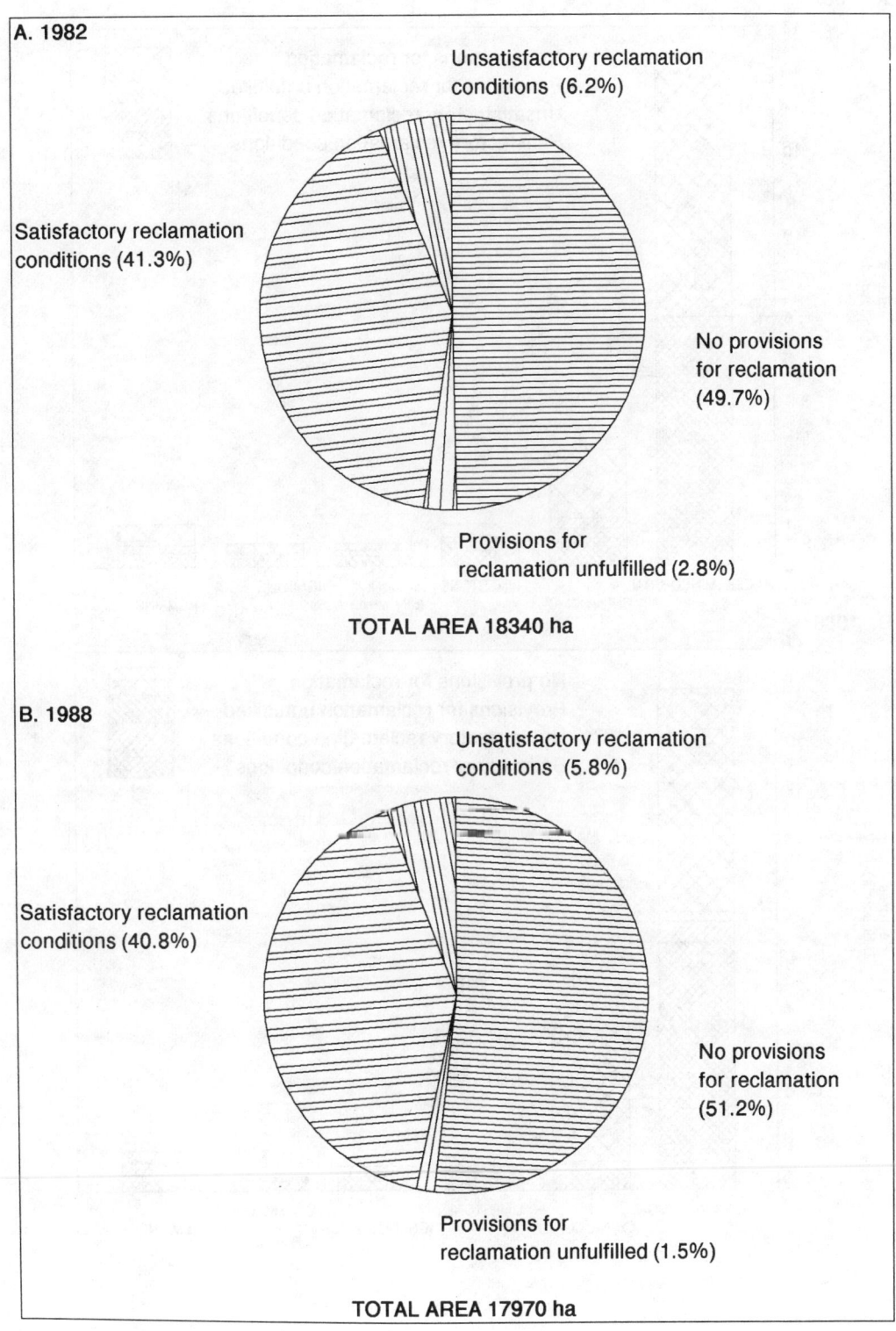

FIGURE 8.3 AREA AND NATURE OF PROVISIONS FOR RECLAMATION FOR THE SURFACE DISPOSAL OF MINERAL WORKING DEPOSITS IN ENGLAND IN 1982 AND 1988 – MAIN MINERAL TYPES.

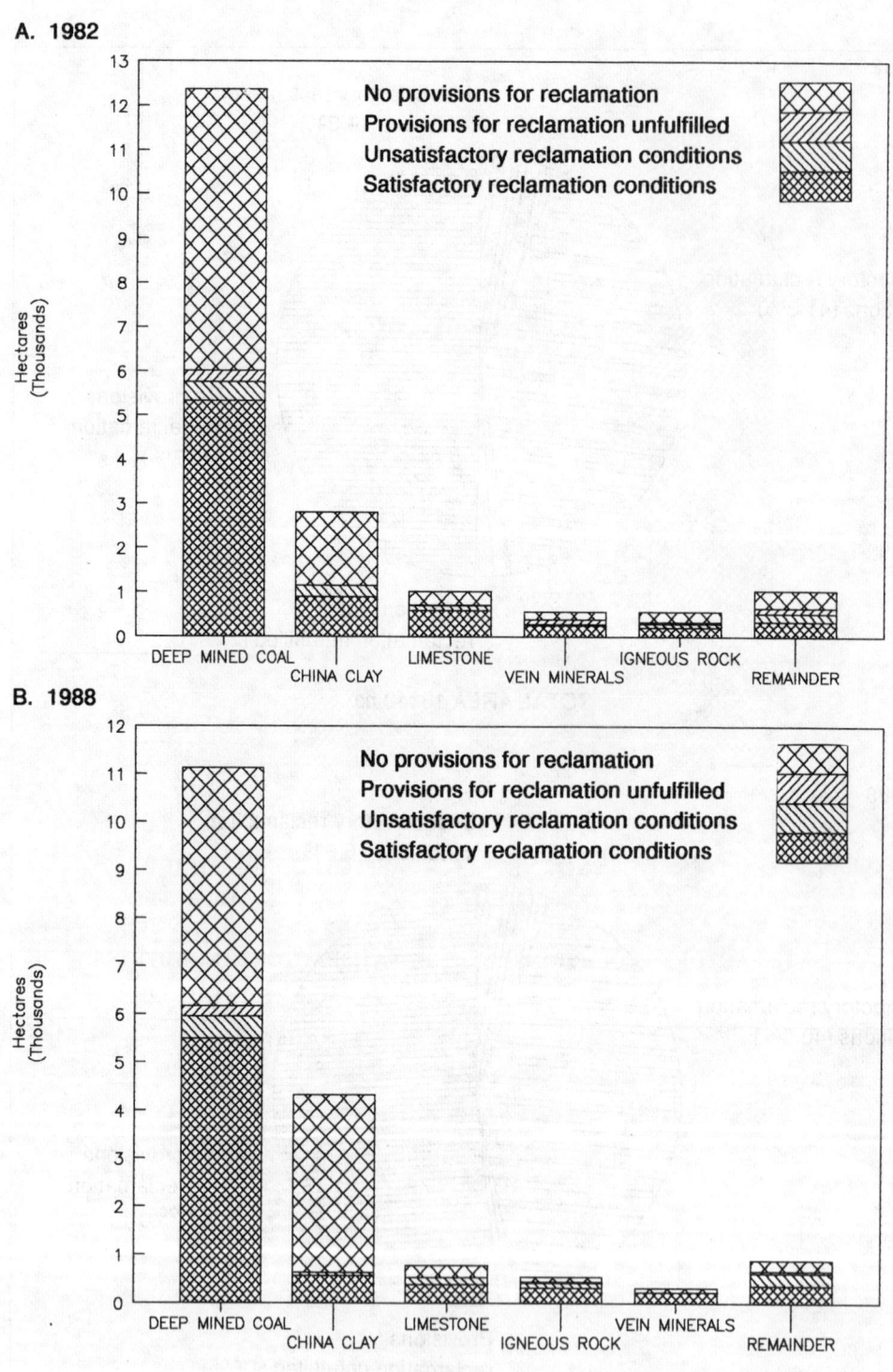

FIGURE 8.4 AREAS FOR THE SURFACE DISPOSAL OF MINERAL WORKING DEPOSITS WITH RECLAMATION CONDITIONS IN 1982 AND 1988 INDICATING NATURE OF CONDITIONS AND AREAS NOT YET AFFECTED.

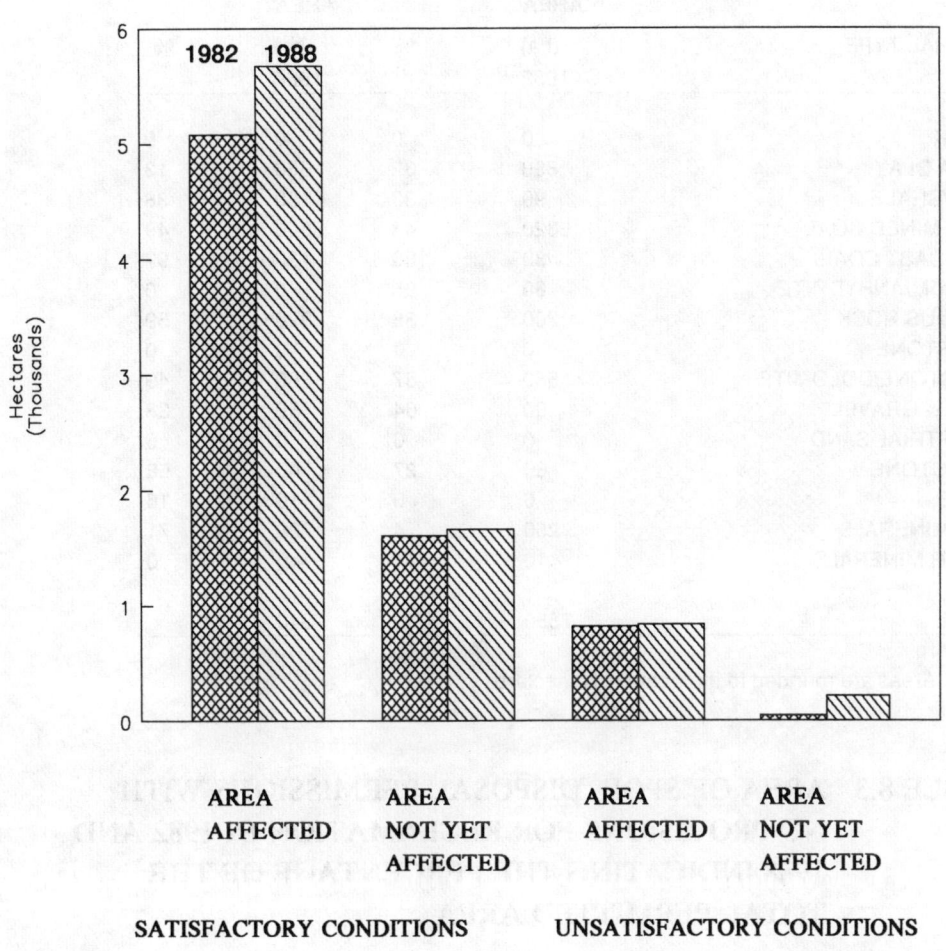

TABLE 8.2 AREA OF SPOIL DISPOSAL PERMISSIONS WITH SATISFACTORY ARRANGEMENTS FOR RECLAMATION IN 1982 AND 1988 INDICATING THE PERCENTAGE OF THE TOTAL PERMITTED AREA

MINERAL TYPE	AREA (ha) 1982	%	AREA (ha) 1988	%
CHALK	0	0	0	0
CHINA CLAY	890	32	560	13
CLAY/SHALE	90	33	130	38
DEEP MINED COAL	5320	43	5500	49
OPENCAST COAL	30	100	130	97
GYPSUM/ANHYDRITE	80	95	0	0
IGNEOUS ROCK	200	36	320	59
IRONSTONE	0	0	0	0
LIMESTONE/DOLOMITE	580	57	380	49
SAND & GRAVEL	80	64	50	33
INDUSTRIAL SAND	0	0	0	0
SANDSTONE	60	27	40	58
SLATE	0	0	10	16
VEIN MINERALS	250	4	220	71
OTHER MINERALS	<10	11	0	0
TOTAL	7580		7340	

Notes: Areas are rounded to the nearest 10 hectares

TABLE 8.3 AREA OF SPOIL DISPOSAL PERMISSIONS WITH NO PROVISIONS FOR RECLAMATION IN 1982 AND 1988 INDICATING THE PERCENTAGE OF THE TOTAL PERMITTED AREA

Hectares

MINERAL TYPE	AREA 1982	%	AREA 1988	%
CHINA CLAY	1650	59	3700	85
CLAY/SHALE	40	15	10	3
DEEP MINED COAL	6530	53	4960	45
OPENCAST COAL	0	0	<10	0
GYPSUM/ANHYDRITE	<10	0	0	0
IGNEOUS ROCK	230	41	100	18
IRONSTONE	100	100	70	100
LIMESTONE/DOLOMITE	310	30	250	32
SAND & GRAVEL	<10	0	20	14
INDUSTRIAL SAND	0	0	10	50
SANDSTONE	150	70	20	33
SLATE	100	61	50	71
VEIN MINERALS	180	32	90	28
OTHER MINERALS	<10	0	30	75
TOTAL	9110		9330	

Note: Areas are rounded to the nearest 10 hectares.

FIGURE 8.5 SURFACE DISPOSAL OF MINERAL WORKING DEPOSITS – MAIN MINERAL TYPES WITH NO PROVISONS FOR RECLAMATION

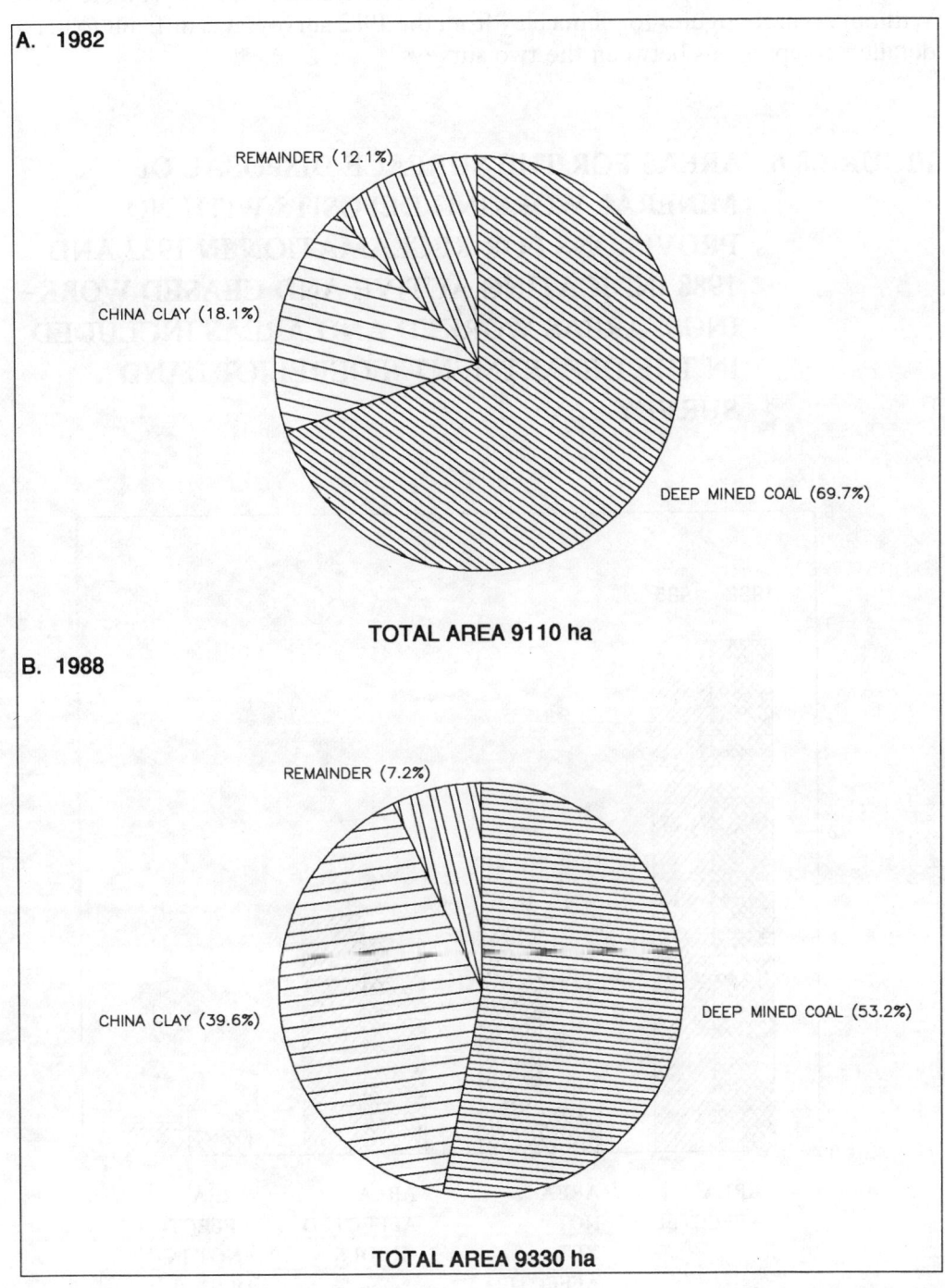

8.11 Almost a quarter of the area with no provisions for reclamation (2200 ha) had not yet been affected. Nearly all of this (2020 ha) was associated with china clay. Half of the area (4580 ha) was affected by active spoil disposal. The remaining 2550 ha were sites where spoil tipping had ceased, of which 88% (2240 ha) were included in the Department's Derelict Land Survey (Figure 8.6). Without accurate figures for china clay from the 1982 survey, it is difficult to make detailed comparisons between the two surveys.

FIGURE 8.6 AREAS FOR THE SURFACE DISPOSAL OF MINERAL WORKING DEPOSITS WITH NO PROVISIONS FOR RECLAMATION IN 1982 AND 1988 INDICATING ACTIVE AND CEASED WORKINGS, AREAS WORKED AND AREAS INCLUDED IN THE DEPARTMENT'S DERELICT LAND SURVEY

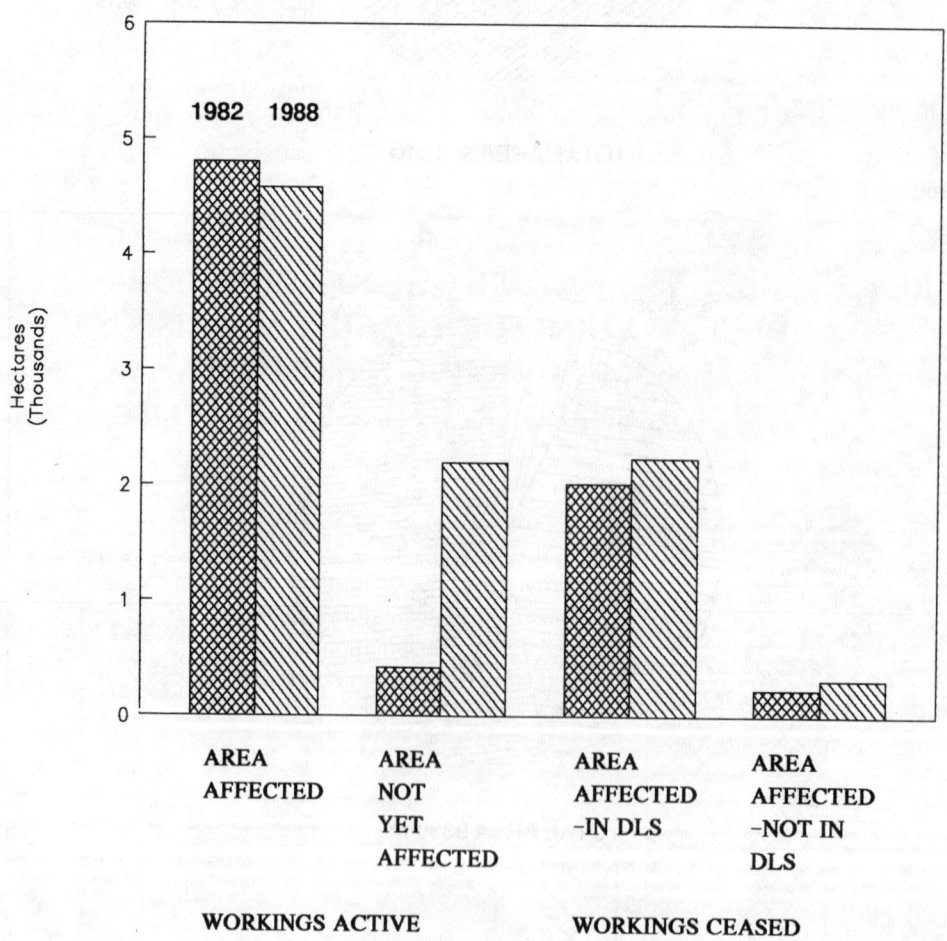

Note: Data for 1982 does not include china clay sites (1650 ha) which were not separately categorised in 1982.

Areas where provisions for reclamation were unfulfilled-

[Information taken from Table 1B of survey return]

8.12 There were 270 ha of spoil tips which were unlikely to be reclaimed because provisions for reclamation were unfulfilled. This was less than 2% all permissions and it represented a reduction of 260 ha since 1982.

8.13 Almost 210 ha (79% of the total) were spoil tips from deep mined coal. Spoil from clay/shale (20 ha) and igneous rock (20 ha) made up of most of the remaining 60 ha - see figure 8.3. Vein minerals, which had accounted for 19% of this category in 1982, were not recorded in 1988.

Regional comparisons

8.14 The South West had the largest area permitted for spoil disposal in England (5260 ha), followed by the East Midlands (4960 ha), and Yorkshire and Humberside (4190 ha). In contrast, the North West had only 790 ha of permissions, the South East 160 ha and none were recorded in East Anglia (Figure 8.7).

8.15 All regions recorded a high proportion of their permitted areas as having no provisions for reclamation. The South West had the greatest problems with 74% of the total (affecting 3880 ha) within this category. Only 20% (1040 ha) of the permitted area in the region had satisfactory reclamation conditions. The Northern region was the next worst affected with 65% of the permitted area having no provisions for reclamation (only 25% had satisfactory reclamation conditions).

FIGURE 8.7 AREA AND NATURE OF PROVISIONS FOR RECLAMATION FOR THE SURFACE DISPOSAL OF MINERAL WORKING DEPOSITS BY STANDARD REGION IN ENGLAND IN 1988.

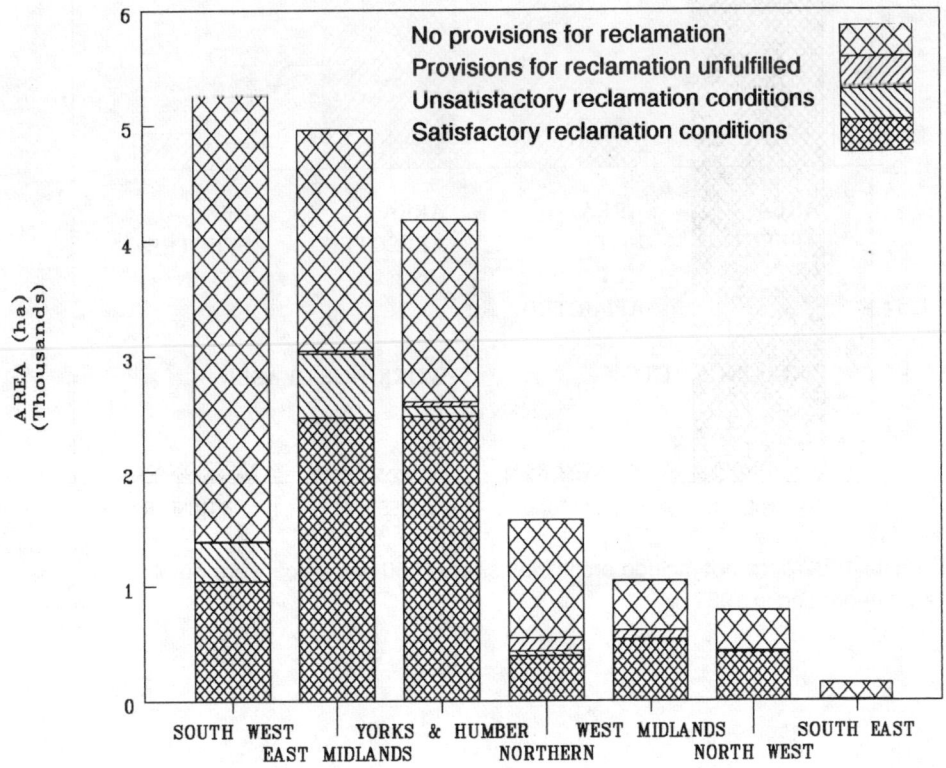

8.16 The East Midlands had the largest total area which included reclamation conditions (3020 ha), equivalent to 61% of the total. However this included a relatively large area with unsatisfactory reclamation conditions (11% of the region's total).

Areas included in the 1988 Derelict Land Survey

[Information taken from Tables 1A and 1B of survey return]
8.17 There were 2460 ha of land used for spoil tips which were also recorded in the Departments' Derelict Land Survey, 40 ha more than in 1982. This represents 14% of the total permitted area for spoil disposal in England. This contrasted with the equivalent figure for surface mineral workings of 2% (see paragraph 7.23).

8.18 Almost the entire area resulted from deep mined coal activities (2280 ha or 93%). Ironstone and limestone accounted for 70 ha and 50 ha respectively (Figure 8.8). Most of the derelict land (2240 ha) resulted from old permissions with no provisions for reclamation.

FIGURE 8.8   LAND DAMAGED BY THE SURFACE DISPOSAL OF MINERAL WORKING DEPOSITS AND INCLUDED IN THE DERELICT LAND SURVEY.

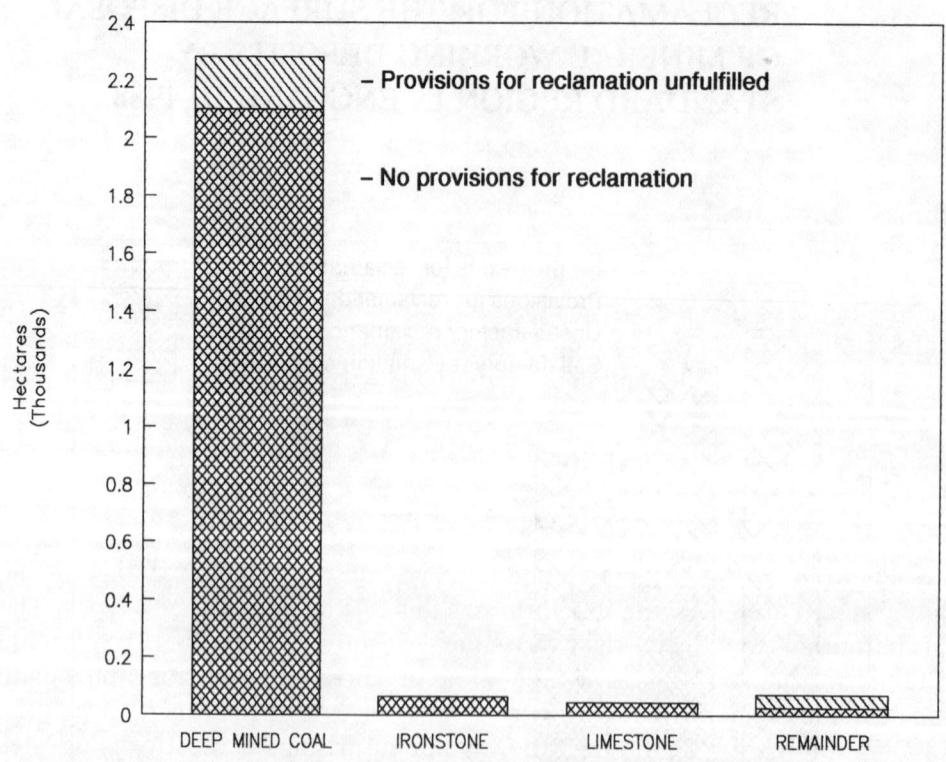

9. LAND COVERED BY PERMISSIONS WHICH INCLUDE AFTERCARE CONDITIONS

Introduction

[Information about aftercare conditions was taken from Table 1D and Table 5 of the survey return and has been collated into Tables C1 to C8 in Volumes 2 and 3 of the survey results].

9.1 Aftercare conditions as defined in the 1981 Act are conditions "requiring that such steps shall be taken as may be necessary to bring land to the required standard for whichever of the following uses is specified in the conditions, namely agriculture, forestry or amenity uses". The power for mpa's to impose such conditions on new permissions has been available since 22nd February 1982. Aftercare conditions can also be imposed as a result of voluntary agreements; and through the making of modifications, discontinuance or prohibition orders in accordance with the provisions of the 1971 Act (Schedules 5 and 9 of the 1990 Act).

9.2 Current planning permissions with aftercare conditions as recorded on Table 1D of the survey form, include sites currently being worked, sites where working has not yet commenced, and sites where reclamation has been completed and aftercare is ongoing. This may include both permissions granted since the implementation of the 1981 Minerals Act and older permissions which have had some form of aftercare conditions imposed through voluntary agreements between the mineral planning authority and the operators. As indicated in Section 14 of this commentary no formal orders have been made as of 1 April 1988 and therefore no aftercare conditions have been imposed through this mechanism. British Coal have had the equivalent of aftercare conditions for opencast sites returning to agriculture since the implementation of the 1958 Opencast Coal Act. These formed the "model" for the wider provisions of the 1981 Minerals Act. The survey does not distinguish between the above individual categories.

9.3 There were 881 permissions which included aftercare conditions (Table 9.1), covering a total area of 21130 ha. Surface mineral workings accounted for 93% (819) of the permissions and 88% of the total area (18600 ha). It should be noted that the survey asked about the number of permissions in respect of aftercare conditions, not the number of sites.

9.4 The majority of land with aftercare conditions - 75% or 15770 ha - had agriculture as the intended final afteruse. This reflects the predominance of agricultural land taken for mineral workings. Almost 3700 ha (17%) were proposed for amenity reclamation and only 1680 ha (8%) for forestry. Figure 9.1, however, indicates differences between surface mineral workings and areas associated with spoil disposal. This is discussed in more detail in paragraphs 9.13 to 9.15.

9.5 For most mineral types in the period covered by the survey (1982 - 1988), the area of land covered by aftercare conditions was only a small percentage of the "total permitted area". Figure 9.2 illustrates this for the main mineral types. The "total permitted area" (ie the maximum which could have had aftercare conditions attached) is taken as the area of current permissions plus land reclaimed under mineral planning permissions since 1982 (as this land may be undergoing aftercare). This observation is not surprising, as most existing sites will be operating under permissions which predate implementation of Section 5 of the 1981 Act.

9.6 Opencast coal had the highest percentage of "total permitted area" covered by aftercare conditions (56%). Equivalent figures for oil/gas (exploration) and sand and gravel were 27% and 21% respectively. Hard rock quarries and other

TABLE 9.1 AREA AND NUMBER OF PERMISSIONS WHICH INCLUDE AFTERCARE CONDITIONS FOR SURFACE WORKINGS AND SPOIL DISPOSAL.

Hectares

MINERAL TYPE	SURFACE MINERAL WORKINGS		SPOIL DISPOSAL	
	AREA	NUMBER OF PERMISSIONS	AREA	NUMBER OF PERMISSIONS
CHALK	290	14	0	0
CHINA CLAY	30	1	350	2
CLAY/SHALE	1120	61	60	1
DEEP MINED COAL	0	0	1620	41
OPENCAST COAL	6930	125	40	4
GYPSUM/ANHYDRITE	130	5	0	0
IGNEOUS ROCK	70	9	140	3
IRONSTONE	10	3	0	0
LIMESTONE/DOLOMITE	1090	50	70	8
OIL/GAS EXPLORATION	50	29	0	0
SAND & GRAVEL	7920	408	0	1
INDUSTRIAL SAND	410	17	0	0
SANDSTONE	130	23	<10	1
SLATE	0	1	<10	0
VEIN MINERALS	190	31	180	1
OTHER MINERALS	250	42	30	0
TOTAL	18590	819	2490	62

Notes: Areas are rounded to the nearest 10 hectares.

mineral types with predominately older, long term permissions had much lower percentages:- limestone 9%, chalk 7%, igneous rock 7%, sandstone 4%, ironstone <1%, slate <1%. This situation can be expected to improve in the future as work on old permissions is completed and new permissions are granted which include aftercare conditions. The mineral review procedure also aims to provide a framework within which mineral planning authorities and the industry can work together to ensure that so far as possible existing mineral operations continue to adapt to meet the standards of the day.

Surface mineral workings

9.7 There were a total of 819 permissions for surface mineral workings which included aftercare conditions, covering an area of 18590 ha.

9.8 All mineral types recorded in the survey were represented, although half of the permissions (408 permissions) were for one mineral type, sand and gravel. Permissions for opencast coal were second most numerous (125) and for clay/shale third (61) (Table 9.1).

9.9 The area of sand and gravel workings covered by aftercare conditions accounted for 43% of the total (7920 ha). Despite the much smaller number of permissions for opencast coal, the area involved accounted for 37% of the total (6930 ha), reflecting the generally larger size of opencast coal sites. Permissions for clay/shale (1120 ha or 6%) and limestone (1090 ha or 6%) with aftercare conditions accounted for most of the remainder (Figure 9.3).

FIGURE 9.1 PROPOSED ENDUSES OF SURFACE MINERAL WORKINGS AND SPOIL DISPOSAL SITES COVERED BY AFTERCARE CONDITIONS IN 1988.

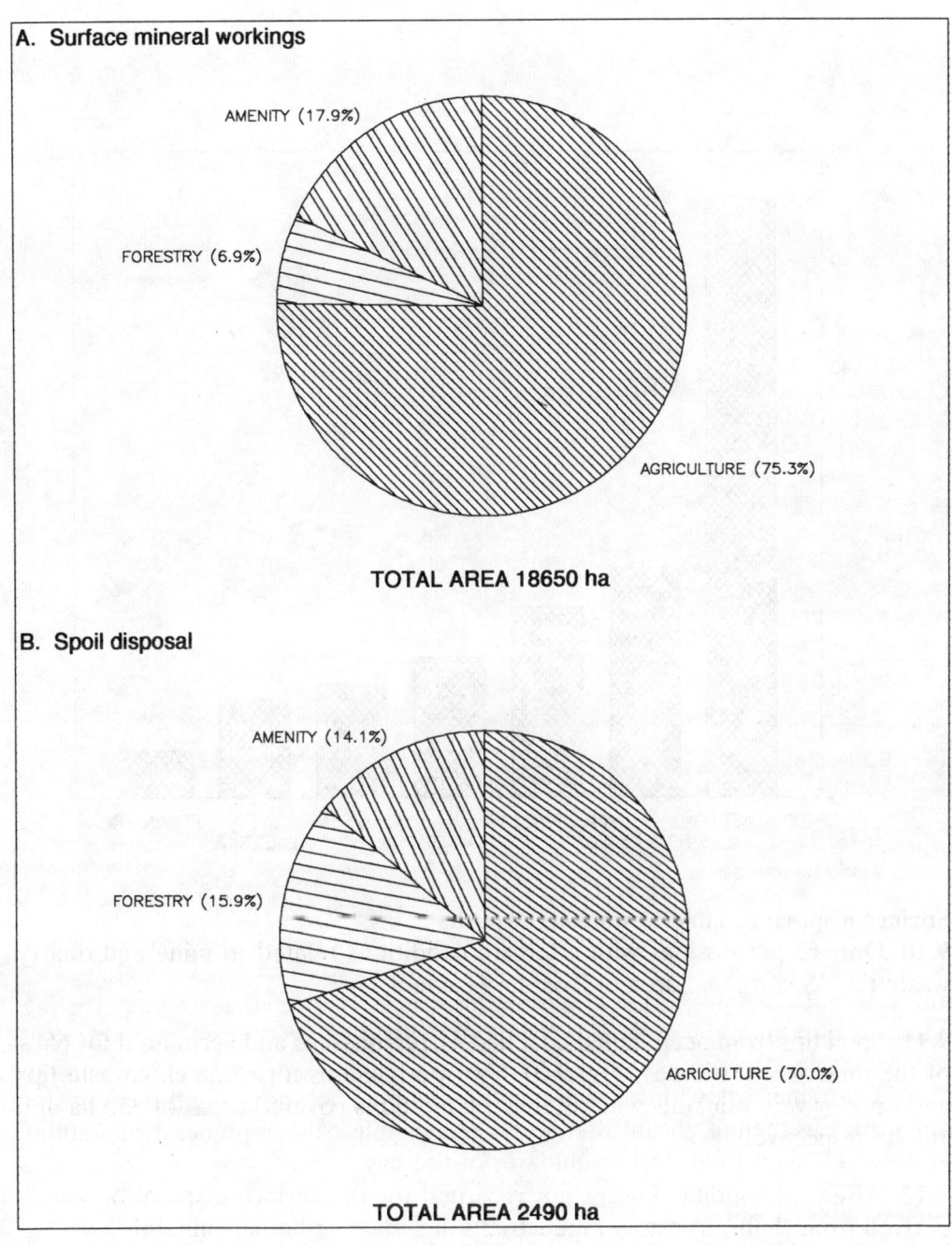

FIGURE 9.2 PERCENTAGE OF THE TOTAL PERMITTED AREA FOR SURFACE MINERAL WORKINGS, SPOIL DISPOSAL AND AREAS RECLAIMED SINCE 1982 WHICH INCLUDE AFTERCARE CONDITIONS.

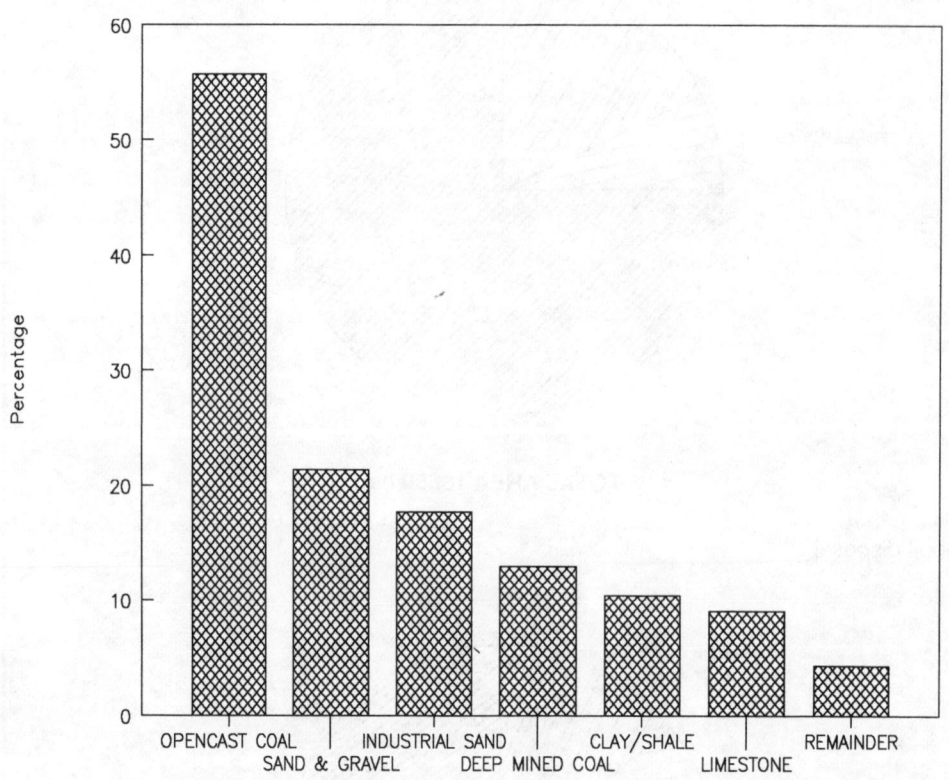

Surface disposal of mineral working deposits

9.10 Only 62 permissions with aftercare conditions related to mine and quarry waste tips, covering an area of 2490 ha (Table 9.1).

9.11 Spoil tips from deep mined coal had 41 permissions and accounted for 65% of the total area (1620 ha) (Fig. 9.3). Two permissions for china clay waste tips and one for vein minerals with aftercare conditions covered areas of 350 ha and 180 ha respectively.

9.12 Aftercare conditions were not recorded for the surface disposal of wastes derived from chalk, ironstone, industrial sand, slate or "other minerals".

Choice of Afteruse

9.13 Over 75% of the area of surface mineral workings which had aftercare conditions were to be reclaimed to agricultre, 18% to amenity and 7% to forestry. Forestry was relatively more common on spoil disposal areas (16%), at the expense of both agriculture and amenity (70% and 14% respectively) (Figure 9.1).

9.14 Figure 9.4 indicates the proposed enduses of the main mineral types. Over 85% of opencast coal workings were to be reclaimed to agriculture, with only 12% to amenity uses and 3% to forestry. In contrast only 65% of sand and gravel sites were to be reclaimed to agriculture, amenity afteruses accounted for 24% and forestry 12%. The proposed afteruses of clay/shale and limestone sites were similar to opencast coal, with over 80% proposed for agricultural uses.

FIGURE 9.3 AREAS OF SURFACE MINERAL WORKINGS AND SPOIL DISPOSAL COVERED BY AFTERCARE CONDITIONS – MAIN MINERAL TYPES.

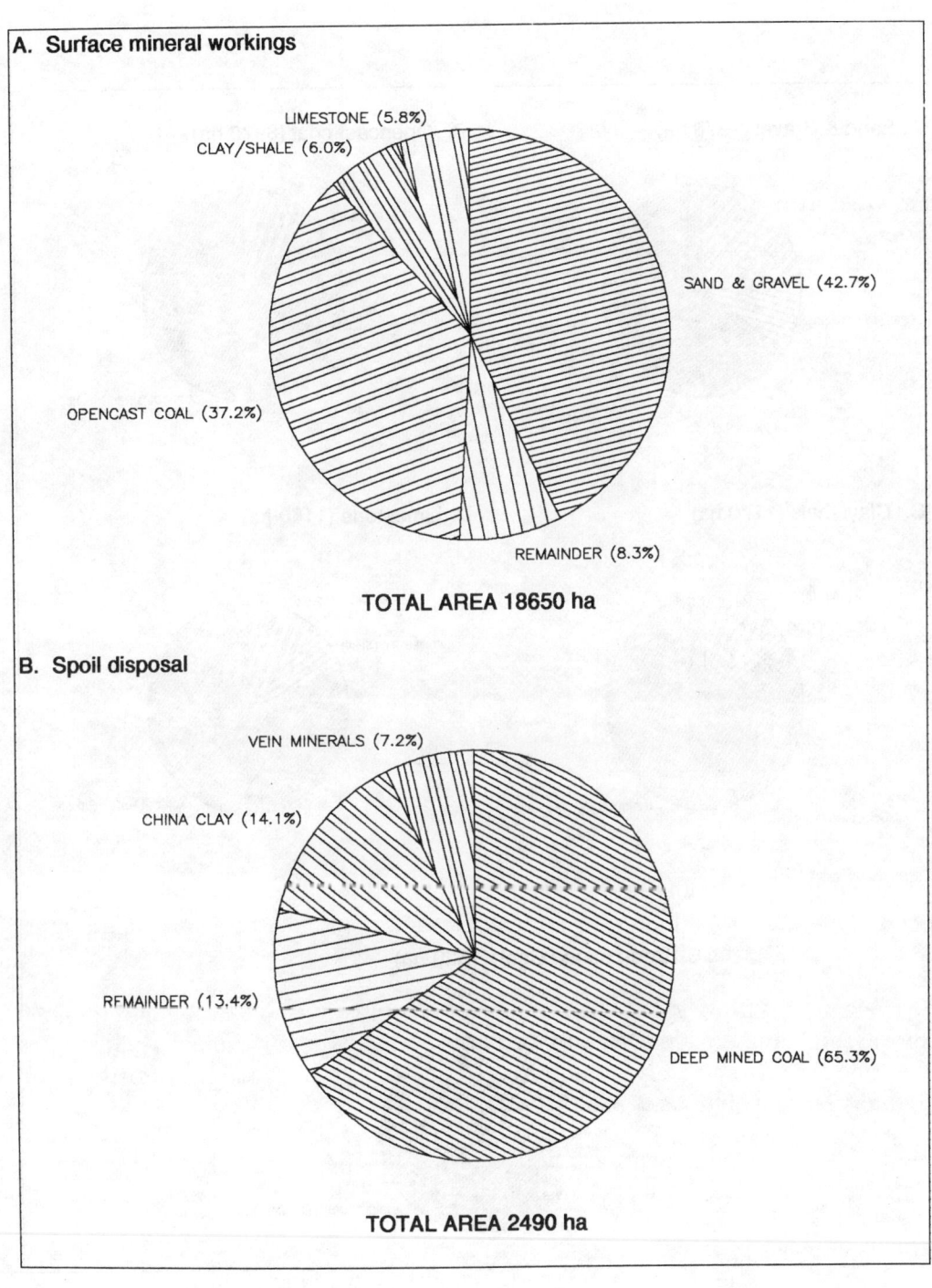

FIGURE 9.4 PROPOSED ENDUSES FOR SURFACE MINERAL WORKINGS AND SPOIL DISPOSAL SITES WITH AFTERCARE CONDITIONS IN ENGLAND IN 1982 – MAIN MINERAL TYPES.

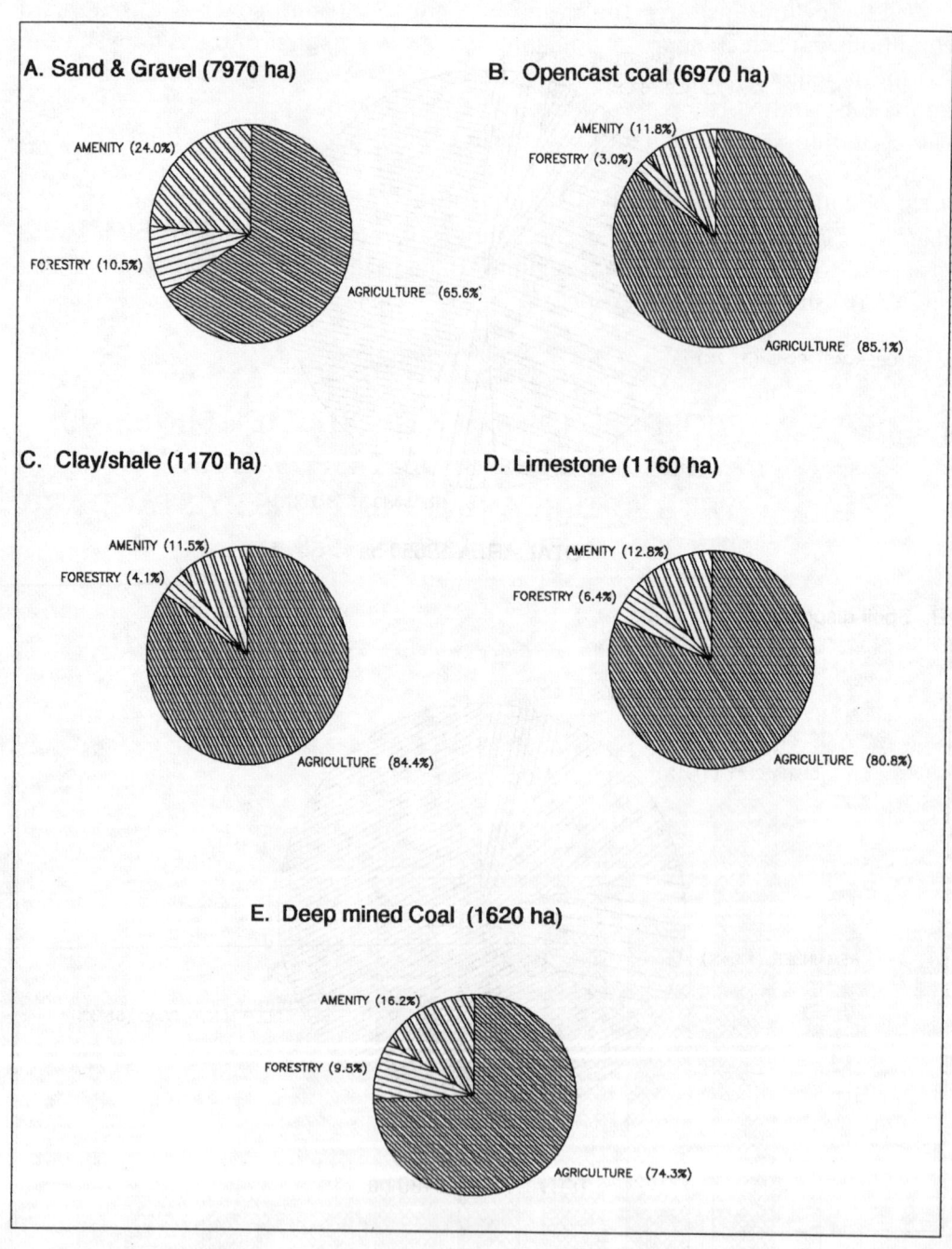

9.15 Agriculture was also the most common enduse proposed for colliery spoil tips (74%). Amenity uses accounted for 16% and forestry (10%).

Regional comparisons

9.16 The East Midlands had the largest area covered by aftercare conditions (4210 ha), followed by the South East (3970 ha) and the Northern region (3490 ha).

9.17 Figure 9.5 illustrates the proposed enduses of land covered by aftercare conditions in each standard region. There were interesting contrasts between the Northern region which had 95% of the area proposed for agriculture and the South East and North West, where only 67% and 48% respectively of the area was chosen for agricultural enduses. Further details are given in sections 15 to 22.

9.18 Significant areas of reclamation to forestry were proposed in only 4 of the regions. The largest area was in the South East (500 ha). A further 450 ha were proposed for the West Midlands region, 310 ha in the South West region and 260 ha in Yorkshire and Humberside.

FIGURE 9.5 PROPOSED ENDUSES OF SURFACE MINERAL WORKINGS AND SPOIL DISPOSAL AREAS WITH AFTERCARE CONDITIONS BY STANDARD REGION IN ENGLAND IN 1988

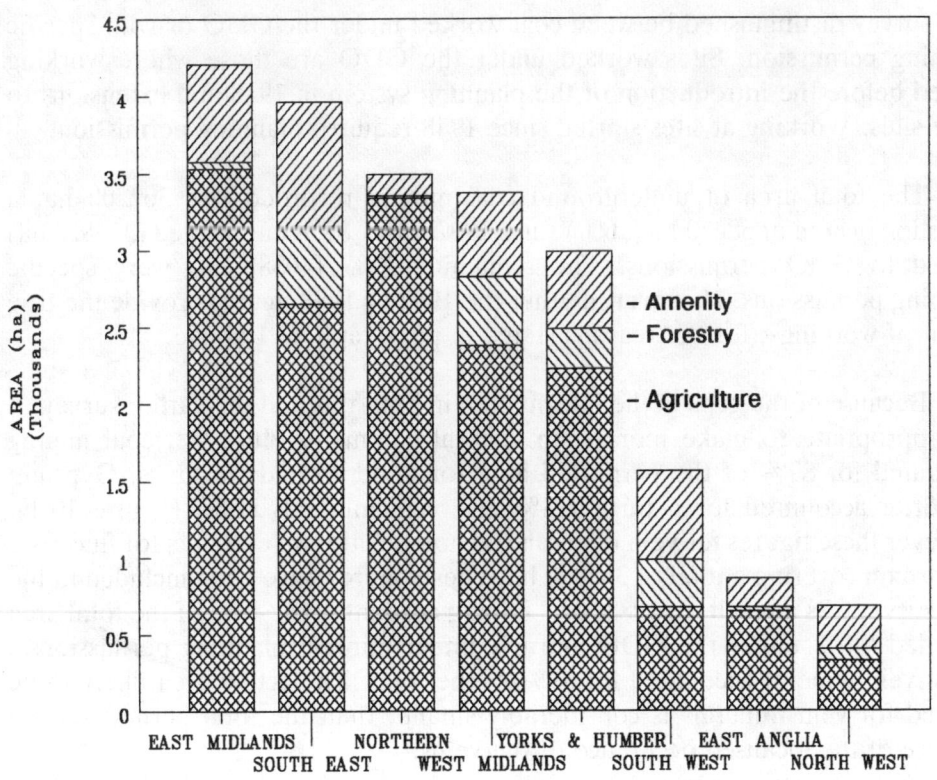

10. PERMISSIONS FOR UNDERGROUND MINING

[Information on underground mining was taken from Table 2 of the survey return and has been collated into Tables D1 to D8 in Volumes 2 and 3 of the survey results].

10.1 Forty five mineral planning authorities recorded underground mining in their area, with total permissions amounting to over 780000 ha. This figure was almost 300000 ha more than recorded in 1982 largely because of a more accurate set of figures for vein mineral permissions compared with 1982. Even so it is acknowledged that the actual figure is higher still, because of the problem of estimating the extent of GDO rights for underground coal mining. However it is unlikely that the full extent of the GDO permissions for underground coal will ever be worked, because in theory these extend to the whole of the coal fields in Great Britain. Neither Durham nor Northumberland, both counties with large areas worked for coal were able to provide information on the extent of GDO sites within their areas in either this or the 1982 survey. Durham were able to provide data on permissions granted since 1982.

10.2 Where more than one mineral was extracted from land covered by a single permission, the mineral that promoted the initial application or development was recorded. The exception to this was where totally different and geologically unrelated minerals were being mined (such as coal from Carboniferous strata beneath gypsum/anhydrite), in which case separate returns were made for each mineral type, and a note was made of the extent of overlapping. In the 1988 survey there were 31480 ha of overlapping. This compares with only 380 ha recorded in 1982.

10.3 In the 1982 survey all coal mining was classified under a single heading. The 1988 survey distinguished between coal worked under the GDO or with specific planning permission. Sites worked under the GDO are those where working started before the introduction of the planning system in 1948 and extensions to those sites. Working at sites started since 1948 requires planning permission.

10.4 The total area of underground coal mining recorded was 362200 ha, a reduction in area of 68470 ha (16%) since 1982. Over 87% of this area (315820 ha) related to GDO permissions. The remaining 13% (46380 ha) were specific planning permissions. However, neither the 1982 or 1988 figures provide the true extent of workings, for the reasons stated in paragraph 10.1 above.

10.5 Because of the gaps in the information in both this and the earlier survey, it is inappropriate to make more than general comments. In 1982, coal mining accounted for 89% of the permitted area for underground workings. Gypsum/anhydrite accounted for a further 4% and vein minerals (2%) (Figure 10.1). However these figures for vein minerals did not include two consents for fluorspar in Durham covering a total of 368000 ha. These figures have been included in the 1988 survey. As a result, in 1988, coal mining accounted for 46% of the total area recorded (40% under the GDO and 6% under mineral planning permissions), whilst vein mineral accounted for 49% of the total. The actual area likely to be worked for vein minerals is considerably smaller than the total permitted area because of the localised occurance of the veins.

10.6 The permitted area for the underground mining of ironstone decreased by 4640 ha (77%) to 1360 ha since 1982 (Table 10.1), whilst gysum/anhydrite permissions decreased by 2050 ha to 14670 ha (12%). In contrast "other minerals" increased by 4540 ha, clay/shale by 370 ha and limestone by 120 ha.

FIGURE 10.1 AREA OF UNDERGROUND MINING, (GDO AND MINERAL PLANNING PERMISSIONS) RECORDED IN ENGLAND IN 1982 AND 1988 – MAIN MINERAL TYPES.

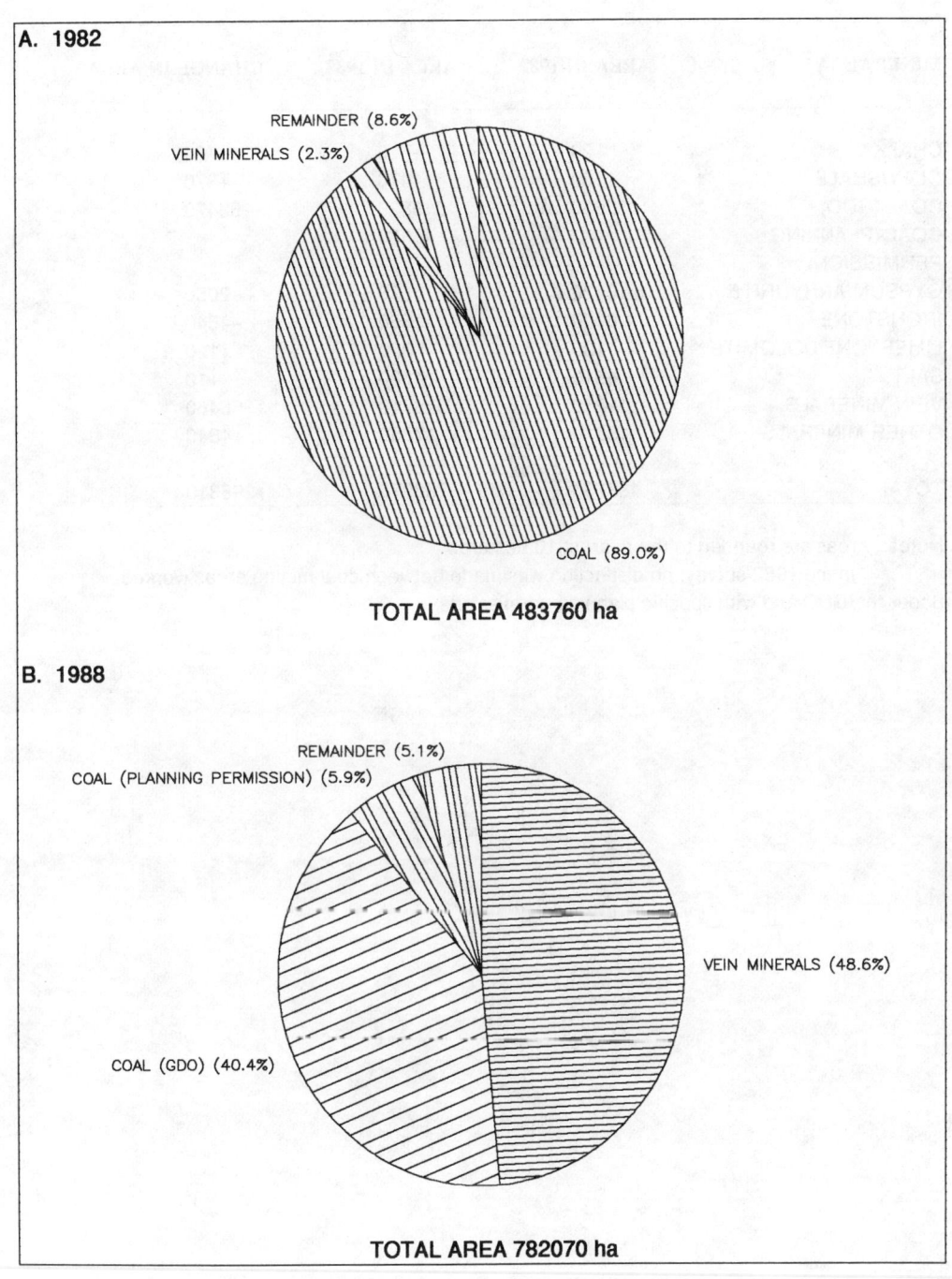

TABLE 10.1 EXTENT OF UNDERGROUND MINING PERMISSIONS BY MINERAL TYPE IN ENGLAND IN 1982 AND 1988

Hectares

MINERAL TYPE	AREA IN 1982	AREA IN 1988	CHANGE IN AREA
CHALK	20	0	−20
CLAY/SHALE	1100	1470	+370
COAL (GDO)	430670	315820	−68470
COAL (PLANNING PERMISSION)		46380	
GYPSUM/ANHYDRITE	16720	14670	−2050
IRONSTONE	6000	1360	−4640
LIMESTONE/DOLOMITE	180	300	+120
SALT	4510	4520	+10
VEIN MINERALS	11250	379710	+368460
OTHER MINERALS	13310	17850	+4540
TOTAL	483760	782070	+298310

Notes: Areas are rounded to the nearest 10 hectares
In the 1982 survey, no distinction was made between coal mining areas worked under the GDO and with specific planning permissions.

11. RECLAMATION OF MINERAL WORKINGS

[Information on reclamation was recorded by mpa's on Tables 4A and 4B of the survey returns. and have been collated into Tables F and G in Volumes 2 and 3 of the survey results.]

Introduction

11.1 The 1988 survey for the first time distinguished between sites reclaimed under planning permission granted for the working of the mineral and those reclaimed under planning permissions for subsequent development or under the Derelict Land Grant Programme, referred to as "other means" below. There was also for the first time a breakdown by afteruse of reclaimed land (agriculture, forestry, amenity and "other uses"). However, to keep the information which DOE sought from mpa's to an acceptable level, the survey did not ask about areas "reclaimed to water" (as requested in 1982). All relevant reclaimed water areas will have been included within the amenity afteruse statistics.

11.2 To provide comparative data with the 1982 survey, the discussion below relates to all reclamation (Tables 4A plus 4B), unless otherwise stated. Even so a direct comparison of the total areas reclaimed in the 1982 and 1988 surveys would still be misleading, as they cover different lengths of time. The 1988 survey recorded reclamation over the six year period 1982–88, compared with the eight year period 1974–82 recorded in 1982. The 1982 survey also recorded the area reclaimed before 1974.

11.3 It was not possible to obtain data for the area of land reclaimed within North Yorkshire or Oxfordshire between 1982 and 1988. These two authorities accounted for 6.3% of all reclamation recorded in the 1982 survey (1780 ha) and would therefore have made an important contribution to the 1988 dataset. For comparative purposes, data for these two counties from the 1982 survey has been taken as the best estimate of later reclamation, and used to assess changes in the rate of reclamation in England (paragraph 11.4). However, it is considered inappropriate to use this data for other comparisons and it has not been included elsewhere.

Area of land reclaimed

11.4 Excluding reclamation in North Yorkshire and Oxfordshire, 20590 ha of mineral working land were recorded as reclaimed during the six year period of this survey, (16900 ha by mineral planning permissions and 3690 ha by "other means"), representing an average annual rate of reclamation of 3430 ha (Table 11.1). This represents a decrease of 3% compared to the period 1974–1982 (average rate of reclamation - 3550 ha/yr). However, if the 1982 data for Yorkshire and Humberside and Oxfordshire are included in the 1988 dataset, the average annual rate of reclamation increases to 3650 ha, representing an increase of 3% since the last survey.

11.5 There was little change in the balance of the main mineral types reclaimed over the period 1982–1988 compared with 1974–1982 as shown in Figures 11.1 and 11.2. Sand and gravel reclamation remained the most common (9370 ha) accounting for 45% of the total. Opencast coal and deep mined coal remained the second and third most commonly reclaimed mineral types accounting for 20% (4080 ha) and 13% (2630 ha) respectively. This contrasts with the situation before 1974, when opencast coal and ironstone had proportionally much larger areas reclaimed (Figure 11.3). The pre 1974 figures for ironstone reflect many sites reclaimed using the Ironstone Restoration Fund. However the domestic extraction of this mineral declined sharply from 1970, ceasing in 1985 when the Fund itself was also wound up.

TABLE 11.1 TOTAL AREA OF LAND RECLAIMED AND AVERAGE ANNUAL RATE OF RECLAMATION IN ENGLAND BETWEEN 1974 – 1982 AND 1982 – 1988. ALL MINERALS.

Hectares

MINERAL TYPE	AREA RECLAIMED 1974–1982 (1)		AREA RECLAIMED 1982–1988 (2)	
	TOTAL AREA ha	ANNUAL RATE ha/yr	TOTAL AREA ha	ANNUAL RATE ha/yr
CHALK	360	46	440	73
CHINA CLAY	–	–	110	18
CLAY/SHALE	2140	268	1360	227
DEEP MINED COAL	3956	495	2630	438
OPENCAST COAL	6180	773	4080	680
GYPSUM/ANHYDRITE	90	11	120	20
IGNEOUS ROCK	110	14	150	24
IRONSTONE	1040	130	530	88
LIMESTONE/DOLOMITE	750	94	620	148
OIL/GAS EXPLORATION	–	–	90	15
OIL/GAS PRODUCTION	–	–	10	<2
SAND & GRAVEL	12410	1551	9370	1561
INDUSTRIAL SAND	430	54	190	31
SANDSTONE	290	37	240	40
SLATE	0	0	<10	<1
VEIN MINERALS	360	46	290	49
OTHER MINERALS	260	32	380	63
TOTAL	28390	3548	20560	3431

Notes: Total areas are rounded to the nearest 10 hectares.
(1) Excludes Hampshire
(2) Excludes North Yorkshire and Oxfordshire.

11.6 Data for the reclamation of land affected by china clay workings was provided for the first time in this survey. A total of over 170 ha were reclaimed between 1982 and 1988.

11.7 Figure 11.4 provides an analysis of the amount of land reclaimed between 1982 and 1988 as a percentage of the current permitted area (surface workings and spoil disposal) from Tables 1A to 1C of the survey (ie total of areas worked and not yet reclaimed, plus unworked permissions). The area of opencast coal workings reclaimed during the six year period of this survey was equivalent to 48% of the current permitted area. Equivalent figures for sand and gravel sites and spoil disposal areas from deep mined coal were 32% and 23% respectively. In contrast figures for slate and china clay were only 1.3% and 2.6% respectively. All hardrock minerals were less than 10%. In part this type of analysis reflects the different working methods between deep hardrock quarries and sand and gravel and opencast sites, where the latter sites are worked and reclaimed more rapidly, often on a progressive basis.

FIGURE 11.1 AREA OF LAND RECLAIMED BETWEEN 1982 AND 1988 INDICATING SATISFACTORY AND UNSATISFACTORY RECLAMATION STANDARDS – MAIN MINERAL TYPES.

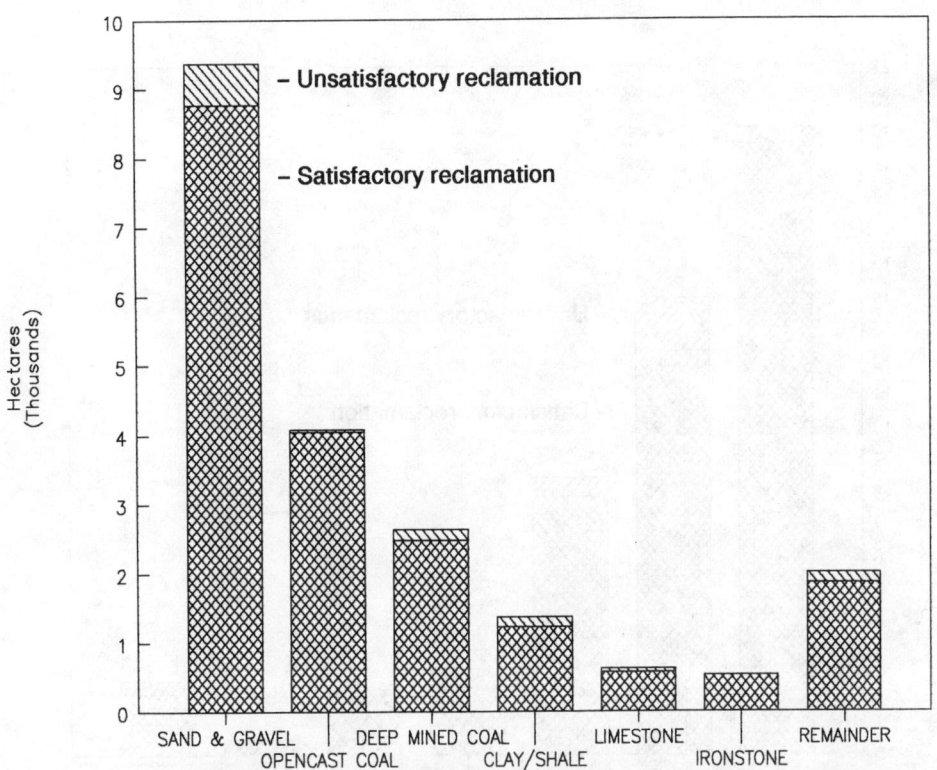

FIGURE 11.2 AREA OF LAND RECLAIMED BETWEEN 1974 AND 1982 INDICATING SATISFACTORY AND UNSATISFACTORY RECLAMATION STANDARDS – MAIN MINERAL TYPES.

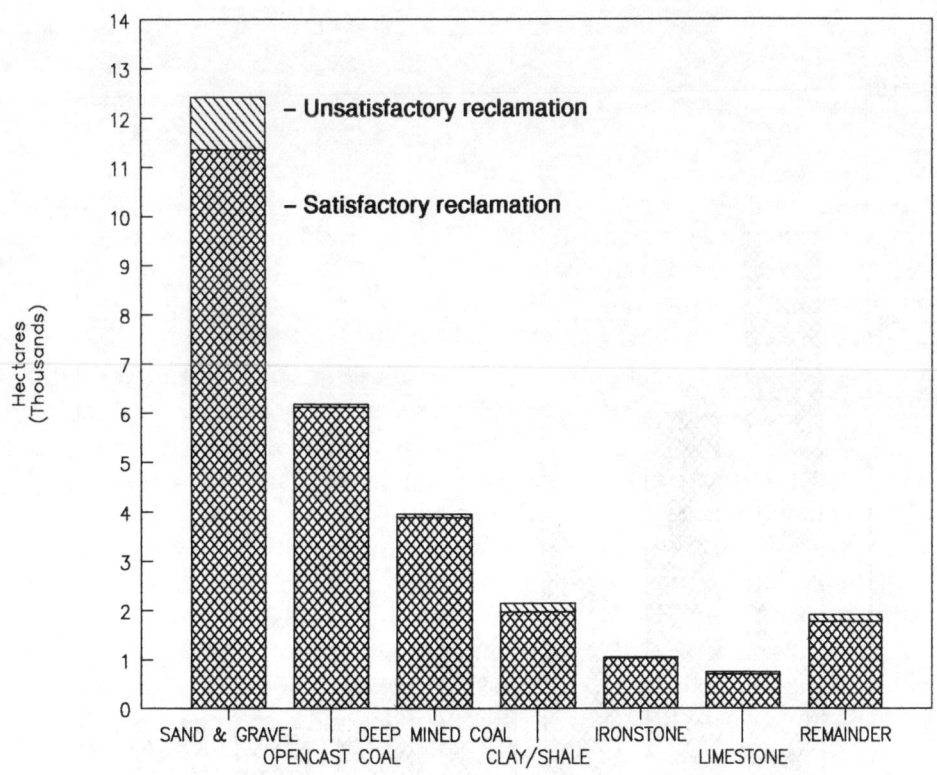

FIGURE 11.3 AREA OF LAND RECLAIMED BEFORE 1974 INDICATING SATISFACTORY AND UNSATISFACTORY RECLAMATION STANDARDS – MAIN MINERAL TYPES.

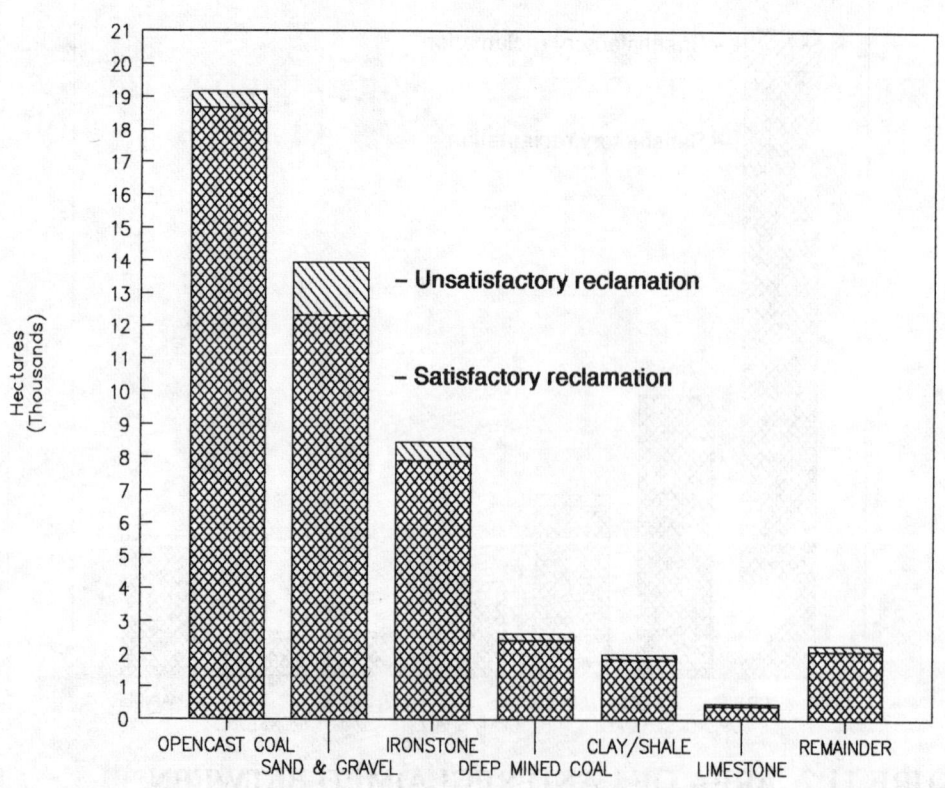

FIGURE 11.4 LAND RECLAIMED BETWEEN 1982 AND 1988 AS A PERCENTAGE OF THE CURRENT PERMITTED AREA – MAIN MINERAL TYPES.

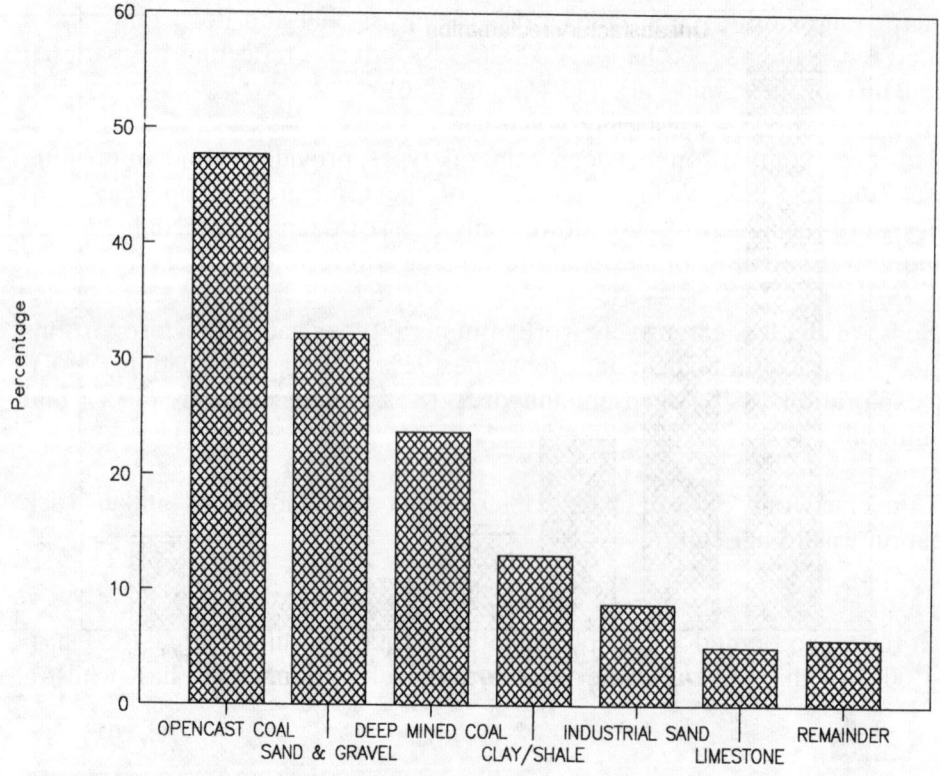

CHOICE OF AFTERUSE

11.8 Mineral working land (surface workings and spoil disposal) was reclaimed to two main afteruses as shown in Figure 11.5. Over 12000 ha (59%) was reclaimed to agriculture and 5720 ha (28%) to amenity uses. Only 820 ha (4%) were reclaimed to forestry. The relative balance of afteruse however varied markedly between mineral types as discussed below.

FIGURE 11.5   AFTERUSES OF LAND RECLAIMED AFTER MINERAL WORKINGS IN ENGLAND BETWEEN 1982 AND 1988.

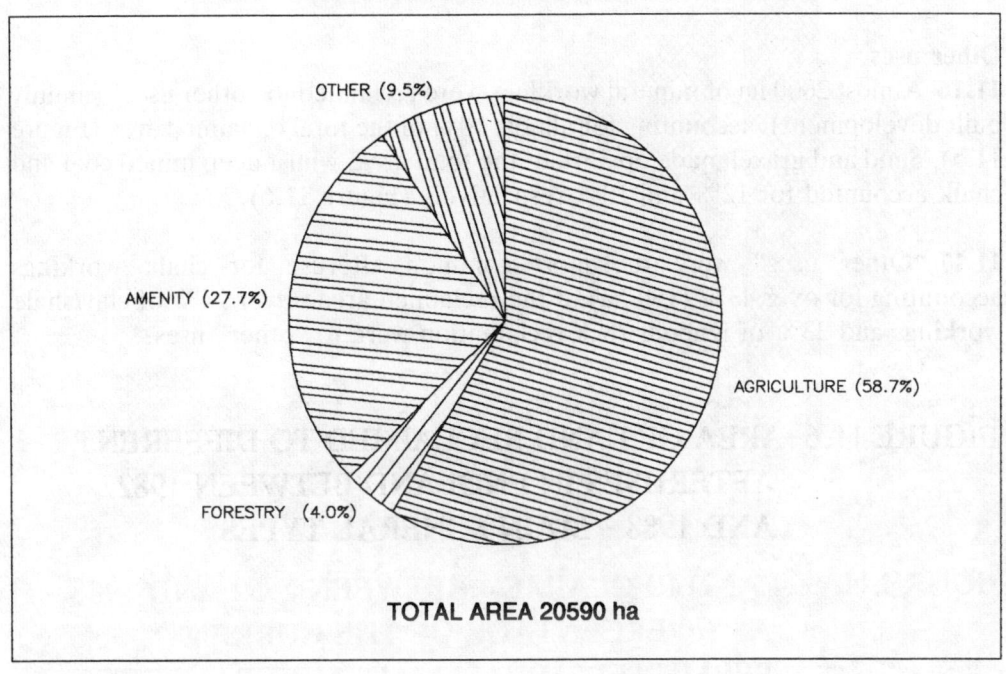

Agriculture

11.9 Three mineral types together made up 80% of the total area of 12090 ha reclaimed to agriculture. Sand and gravel, opencast coal and deep mined coal accounted for 4840 ha, 3360 ha and 1420 ha respectively, reflecting the national predominance of these mineral types (Figure 11.6).

11.10 However comparison between mineral types provided some interesting contrasts (Figure 11.7). Whilst over 80% of the total area of opencast coal workings were reclaimed to agriculture, only 54% of deep mined coal, 52% of sand and gravel and 38% of clay/shale reclamation were to this afteruse.

11.11 Some of the less extensively worked mineral types had higher proportions reclaimed to agriculture, including ironstone (92%), oil/gas production (90%), oil/gas exploration (89%), gypsum/anhydrite (82%), vein minerals (79%) and other minerals (74%).

11.12 Conversely only 38% of chalk, 21% of china clay, and 7% of igneous rock reclamation was to agriculture.

Forestry

11.13 Forestry accounted for only 4% of the total reclaimed area (820 ha). Almost 60% of this was on former sand and gravel workings (480 ha), whilst a further 22% was on spoil disposal areas associated with deep mined coal (190 ha) - Figure 11.6.

Amenity

11.14 Amenity reclamation accounted for 28% (5720 ha) of reclaimed land, the second most common afteruse (Figure 11.5). With the exception of oil/gas (exploration and production) all mineral types recorded had some area reclaimed to some form of amenity afteruse. By far the most extensive mineral with this afteruse was sand and gravel (3343 ha), 58% of all amenity reclamation (Figure 11.6). Other minerals with large areas reclaimed to amenity uses were deep mined coal tips (790 ha), opencast coal (540 ha) and clay/shale (460 ha).

11.15 Amenity reclamation was the most common afteruse for china clay (53%), igneous rock (63%) and industrial sand (42%). In contrast, only 13% of opencast coal and 5% of ironstone were reclaimed to this afteruse.

Other uses

11.16 Almost 2000 ha of mineral workings were reclaimed to "other uses" (mainly built development), accounting for almost 10% of the total reclaimed area (Figure 11.5). Sand and gravel made up 36% of the total area, whilst deep mined coal and chalk accounted for 12% and 10% respectively (Figure 11.6).

11.17 "Other uses" were the most common afteruse for chalk workings accounting for over 43% (190 ha) of the reclaimed area. Over 24% of clay/shale workings and 23% of igneous rock reclamation were to "other" uses.

FIGURE 11.6 AREA OF LAND RECLAIMED TO DIFFERENT AFTERUSES IN ENGLAND BETWEEN 1982 AND 1988 – MAIN MINERAL TYPES.

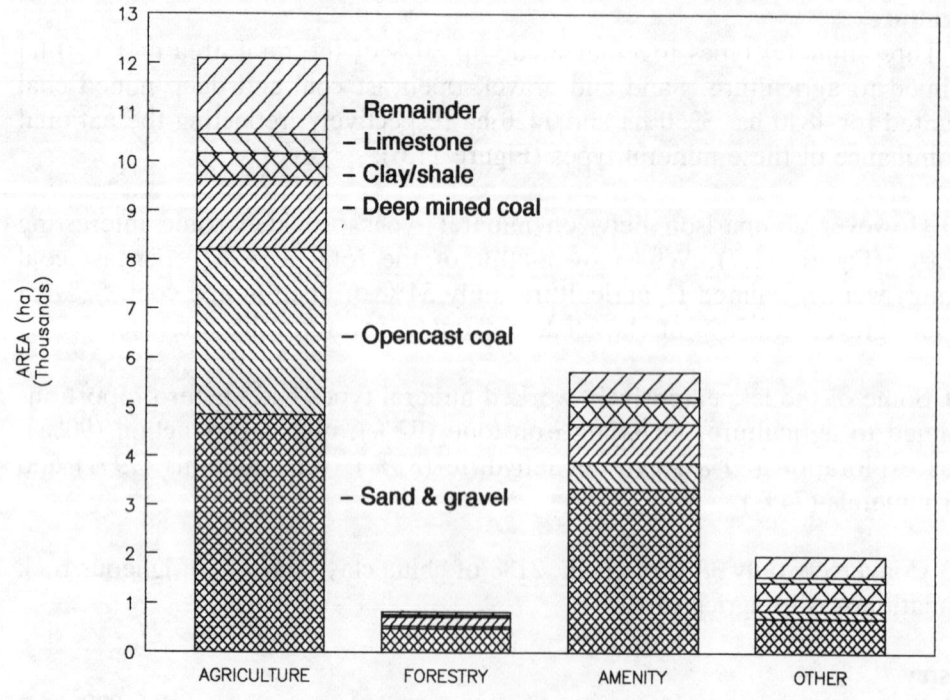

FIGURE 11.7 AFTERUSES OF LAND RECLAIMED BETWEEN 1982 AND 1988, FOLLOWING MINERAL WORKING – MAIN MINERAL TYPES.

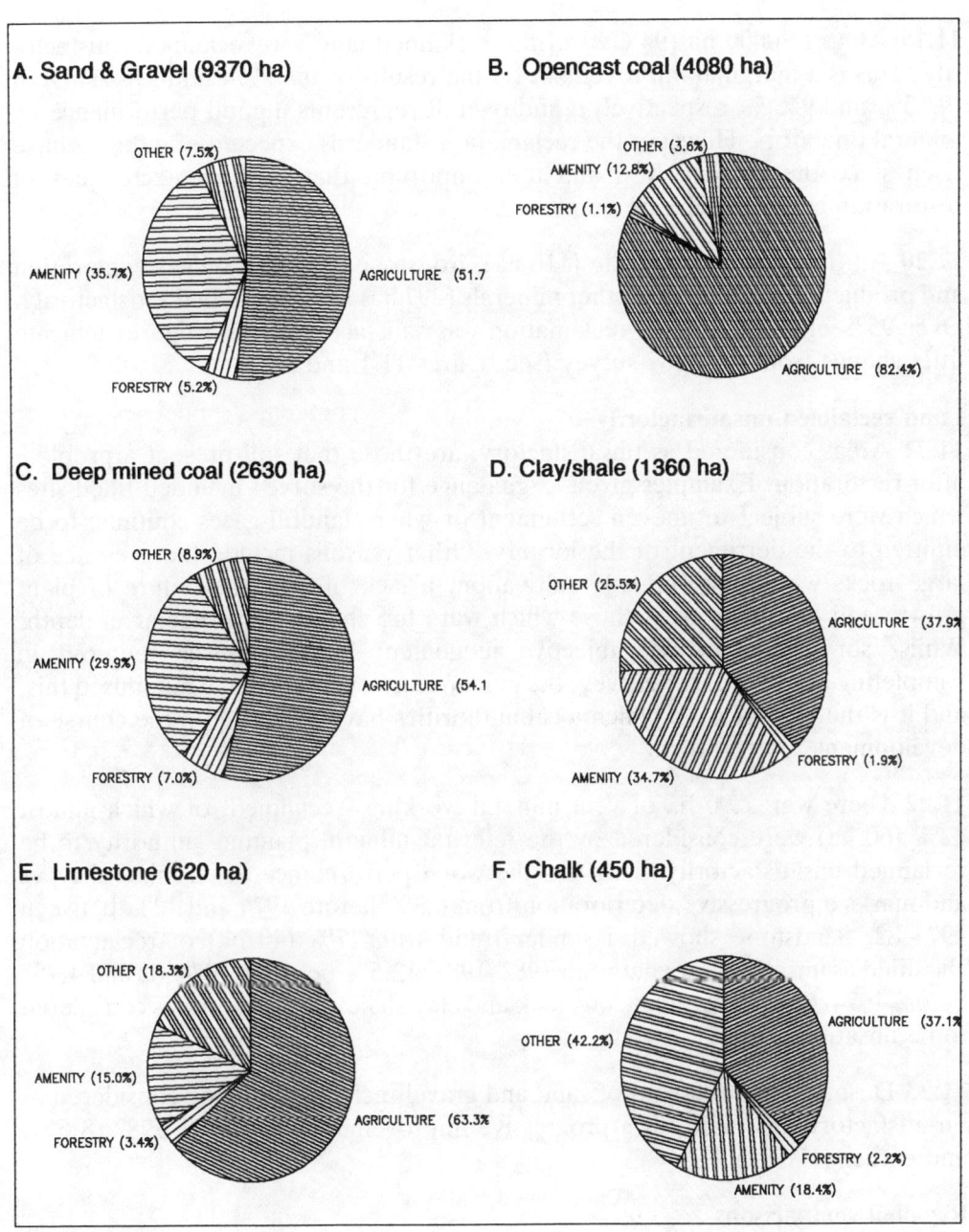

Land reclaimed satisfactorily

11.18 The term "satisfactory reclamation" has been used in this survey to indicate areas that have been reclaimed and do not present a problem for present or future use. It does not necessarily indicate that the sites were reclaimed to their preworking standard or use.

11.19 Almost 20000 ha (94.6%) of the reclaimed land was reclaimed satisfactorily. This is a marginal improvement on the results of the 1974 and 1982 surveys (93.3% and 94.4% respectively); and overall represents a good performance by mineral operators. However the reclamation standards expected of industry have risen since the early 1970's, and it is important that effective techniques of restoration and aftercare are employed.

11.20 All the gypsum/anhydrite (116 ha), ironstone (524 ha), oil/gas exploration and production (101 ha) and other minerals (297 ha) were reclaimed satisfactorily. Over 95% of opencast coal reclamation was satisfactory. These figures indicate little change from the 1982 survey (see figures 11.1 and 11.2).

Land reclaimed unsatisfactorily

11.21 Areas considered as unsatisfactory, are those that still present a problem after restoration. Examples given as guidance for the survey included filled sites which were subject to uneven settlement or where landfill gases continue to be emitted to the detriment of the locality. Other reasons include the presence of large rocks which may impede cultivation, a lack of soil, the failure of plant growth and for water areas those which were too shallow or irregular in depth. Whilst some element of subjective judgement by mpa's was required in completing this part of the survey, the guidance notes should have minimised this, and it is the type of issue which local authorities have to decide in the course of development control work.

11.22 There were 290 ha of vein mineral workings reclaimed, of which almost 22% (60 ha) were considered by the relevant mineral planning authority to be reclaimed unsatisfactorily. This was the worst performance of all mineral types and marks a progressive deterioration from 6.3% before 1974 and 12% between 1974–82. Sandstone showed a similar trend with 17% (40 ha) of reclamation classified as unsatisfactory between 1982–1988, 12.5% between 1974–82 and 4.7% before 1974. Over 10% of igneous rock and clay/shale reclamation was considered to be unsatisfactory.

11.23 Despite 6.4% (600 ha) of sand and gravel reclamation being considered as unsatisfactory this indicates a progressive improvement since 1974–1982 (8.6%) and earlier (11.7%).

Regional comparisons

11.24 The East Midlands recorded the largest area of mineral workings reclaimed between 1982 and 1988 (4850 ha). The South East (4500 ha) and Northern region (3500 ha) had the second and third largest areas reclaimed. East Anglia and the South West recorded the smallest areas of reclamation (1100 ha and 990 ha respectively), see Figure 11.8. These overall regional performances reflect, in part, the general stock of permissions and the range and level of mineral extraction in the areas concerned. The East Midlands has the largest total area for surface mineral workings (Figure 7.6) and the second largest area for spoil disposal (Figure 8.7). It also contains a wide range of minerals. In contrast East Anglia has the smallest total of areas affected by current permissions and these are predominantly for sand and gravel working. However the regional reclamation comparisons also reflect the extent of sites with or without satisfactory reclamation conditions. The South West had the largest total permitted area of spoil tips (Figure 8.7) and the fourth largest of surface mineral workings (Figure 7.6); but the majority of permissions had no provisions for reclamation.

FIGURE 11.8 CHOICE OF AFTERUSE FOR LAND RECLAIMED BETWEEN 1982 AND 1988 BY STANDARD REGION.

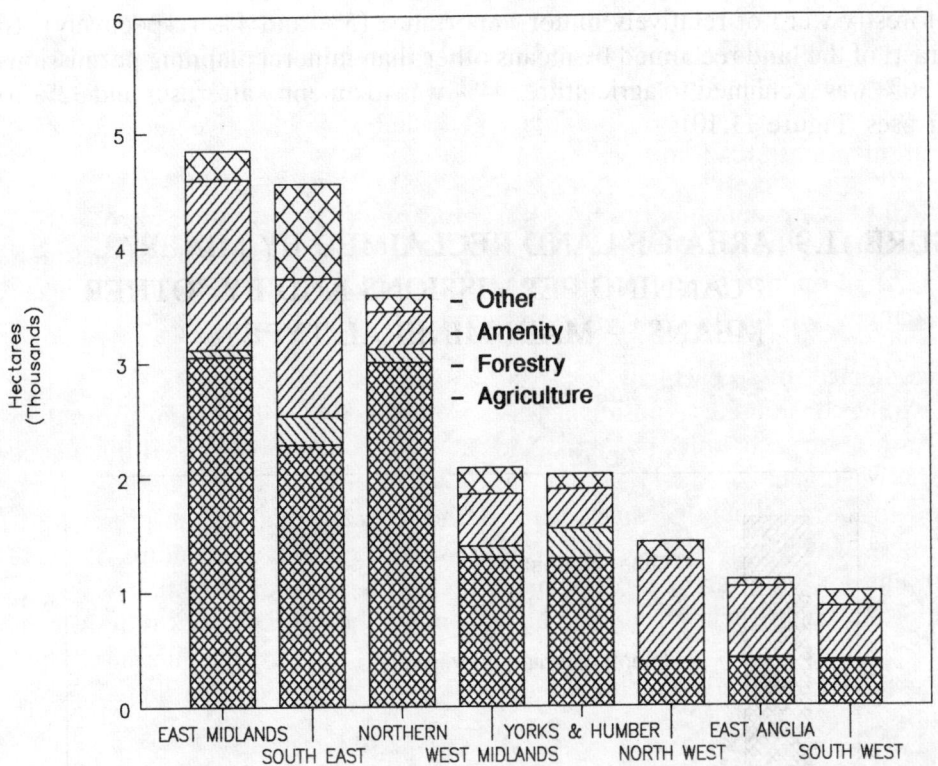

11.25 Agriculture was generally the most common afteruse within each region. The exceptions were the three regions with the smallest areas reclaimed; the North West, East Anglia and the South West, where amenity reclamation was more common. The Northern region had the largest area reclaimed to agriculture, accounting for 86% of all reclamation in the region.

11.26 Amenity reclamation was the second most common afteruse in all regions except within the three mentioned previously. Forestry was of minor significance in all regions except the South East and the West Midlands, whilst "other uses" were particularly significant in the South East. These statistics are discussed more fully under the relevant regional section.

Arrangements through which reclamation was achieved
11.27 Data on land reclaimed between 1982 and 1988 under the mineral planning permissions and by other permissions or with Derelict Land Grant (referred to as "other means") were taken from Tables 4A and 4B respectively.

11.28 The great majority of the reclamation of mineral workings and spoil tips (81% or 16900 ha) was a result of reclamation conditions attached to the relevant mineral planning permissions. The remaining 3690 ha was reclaimed by "other means".

11.29 The mineral types with the largest areas reclaimed by "other means" were spoil tips from deep mined coal (1220 ha), sand and gravel (1140 ha), clay/shale (495 ha), chalk (200 ha) and limestone (150 ha). For deep mined coal sites and chalk this represented 46% and 45% respectively of all reclamation. Clay/shale and limestone also had sizable proportions reclaimed by means other than the mineral planning permissions (38% and 23% respectively) (Figure 11.9). It should be noted that the reclamation of derelict colliery spoil tips has been one of the priority programmes in the Departments Derelict Land Grant programme for most of the relevant period, following the 1981 report "Coal and the Environment" by the Commission on Energy and the Environment (CENE).

11.30 Under mineral planning permissions, 65% of reclamation was to agriculture. Amenity reclamation accounted for a further 26%. Other development and forestry were of relatively minor importance (5% and 4% respectively). In contrast, of the land reclaimed by means other than mineral planning permissions only 30% was reclaimed to agriculture, 34% was to amenity afteruses and 32% to other uses (Figure 11.10).

FIGURE 11.9 AREA OF LAND RECLAIMED BY MINERAL PLANNING PERMISSIONS AND BY "OTHER MEANS" – MAIN MINERAL TYPES.

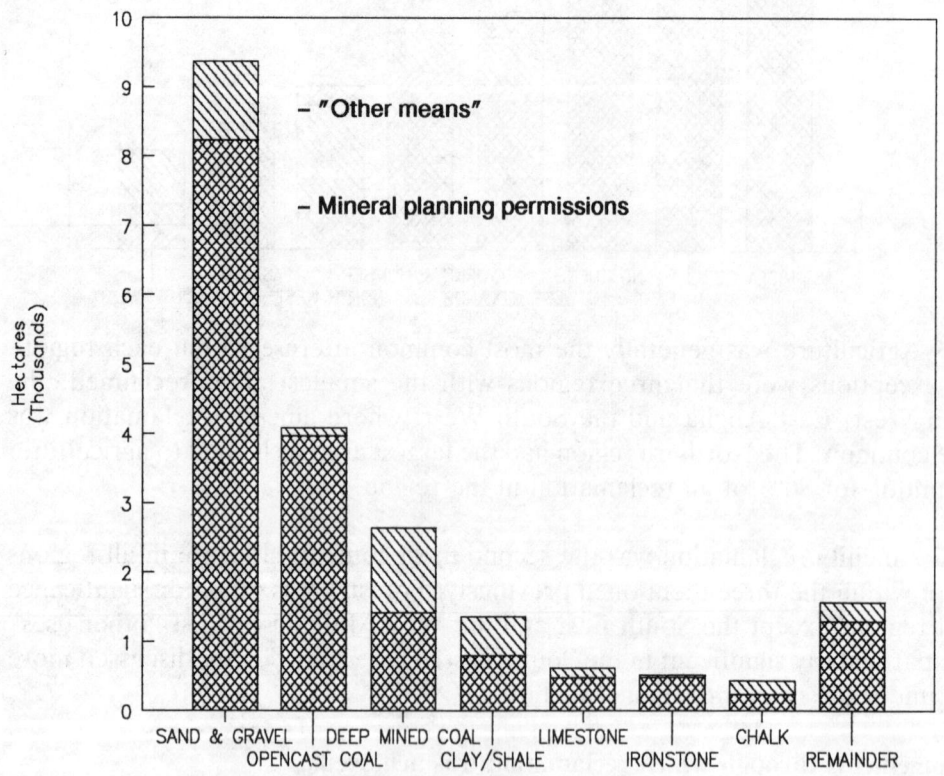

FIGURE 11.10 AFTERUSE OF LAND RECLAIMED BY MINERAL PLANNING PERMISSIONS AND BY "OTHER MEANS" IN ENGLAND BETWEEN 1982 AND 1988

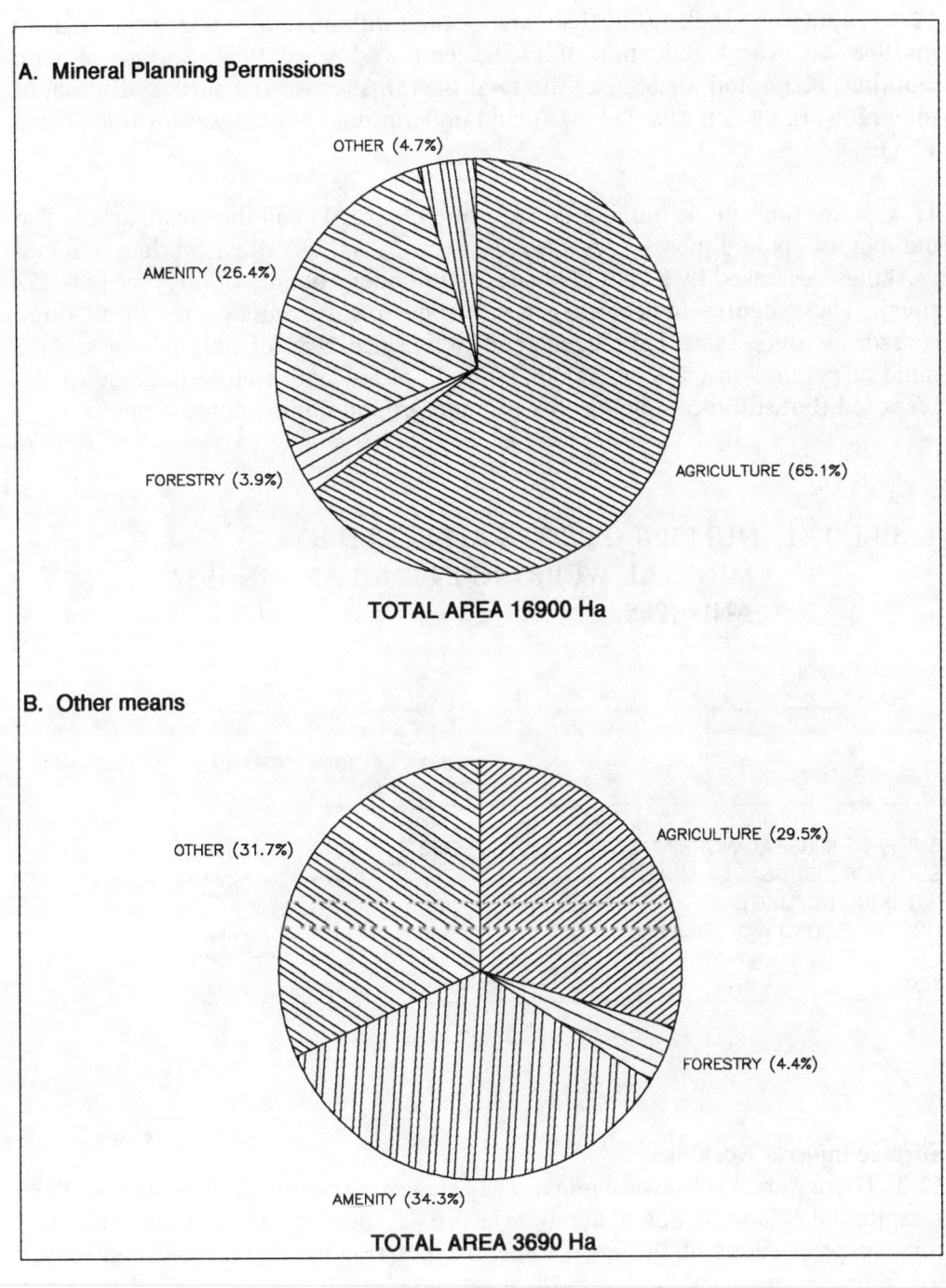

12. NUMBER OF SITES

Introduction

[Figures were taken from Table 5 of the survey return and results have been collated into Tables H1 to H8 of Volumes 2 and 3 of the survey results].

12.1 There were a total of 5131 sites associated with mineral workings in England in 1988, an overall reduction of 1734 sites (25%) since 1982. Surface mineral workings accounted for 85% of the total (4343), sites for the surface disposal of mineral working deposits 9% (474), and underground workings 6% (314) (Table 12.1).

12.2 Reductions in the number of sites were recorded in all three categories. The number of spoil disposal sites decreased by 38% (292 sites), surface mineral workings decreased by 24% (1390 sites) and underground workings by 14% (52 sites). These figures indicate a general trend towards larger sites in all three categories, since there have been reductions since 1982 of only 8% in surface mineral permissions, 2% in "spoil disposal" areas; and a large increase in the recorded (but still incomplete) area of underground mining permissions.

TABLE 12.1 NUMBER OF SITES AFFECTED BY MINERAL WORKING IN ENGLAND IN 1982 AND 1988.

	1982	1988	CHANGE
SURFACE MINERAL WORKINGS	5733	4343	−1390
SURFACE DISPOSAL OF MINERAL WORKING DEPOSITS	766	474	−292
UNDERGROUND WORKINGS	366	314	−52
TOTAL	6865	5131	−1734

Surface mineral workings

12.3 There were 4343 surface mineral working sites recorded in England in 1988. Despite the 24% reduction in number since 1982, there was little difference to the relative proportions of the main mineral types (Figure 12.1). Sand and gravel workings remained by far the most numerous, accounting for 35% of the total (1533 sites). Limestone quarries were the second most common, accounting for 12% (521 sites) and clay/shale third (11% or 496 sites). Sandstone (350 sites), chalk (220 sites) and opencast coal (163 sites) sites accounted for a further 8%, 5% and 4% respectively.

12.4 The number of workings associated with all mineral types except china clay, gypsum/ anhydrite, slate and "other minerals" were reduced (Table 12.2).

12.5 Sand and gravel workings decreased by the largest number, a net reduction of 836 sites (a reduction of 35%). Over 75% of this reduction (631 sites) was due to the closure or amalgamation of sites in the South East region.

FIGURE 12.1 NUMBER OF SITES FOR SURFACE MINERAL WORKINGS IN ENGLAND IN 1982 AND 1988 – MAIN MINERAL TYPES

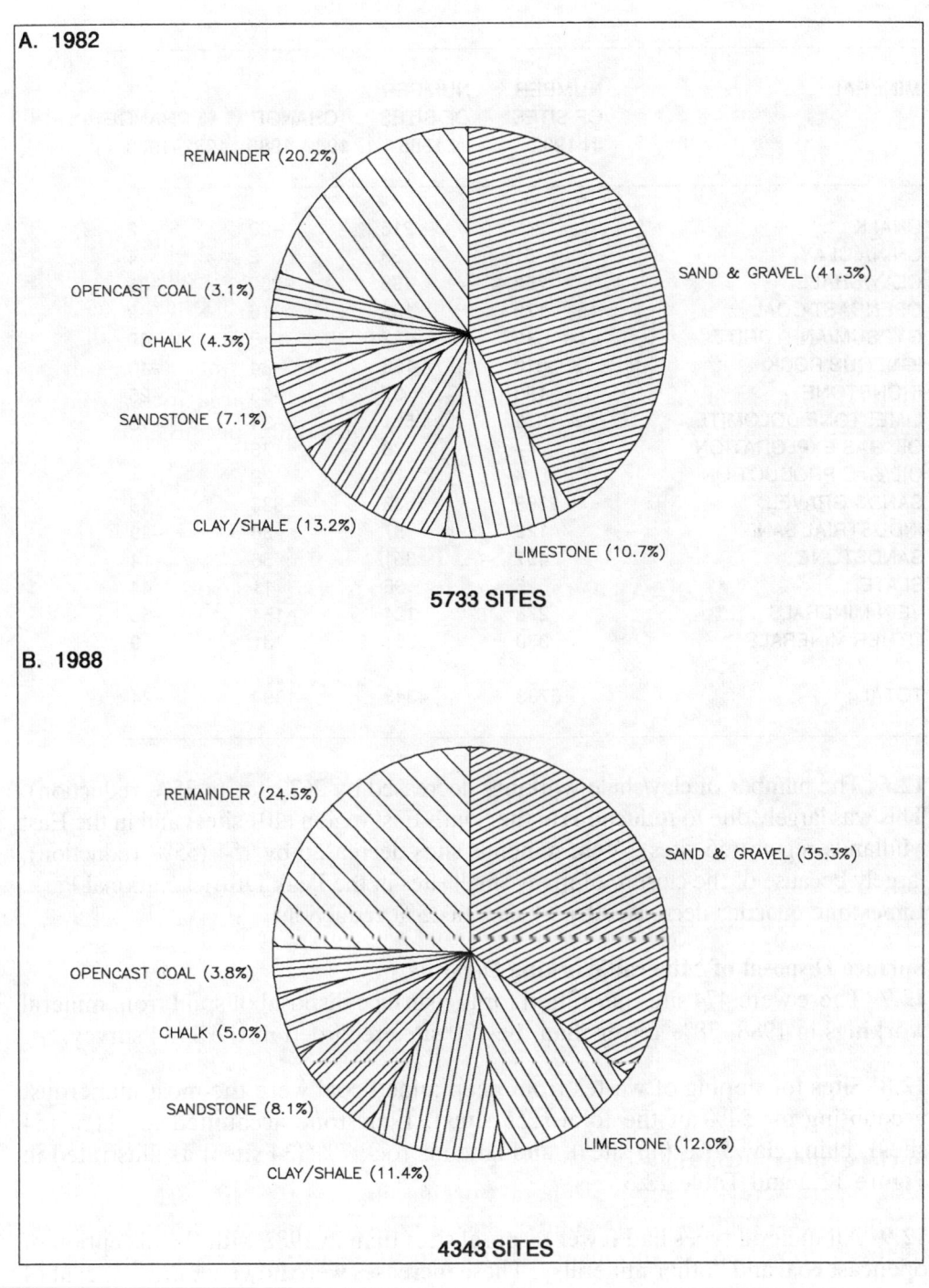

TABLE 12.2 NUMBER OF SURFACE MINERAL WORKINGS IN ENGLAND IN 1982 AND 1988 INDICATING CHANGE IN NUMBER AND PERCENTAGE CHANGE – ALL MINERALS.

MINERAL	NUMBER OF SITES IN 1982	NUMBER OF SITES IN 1988	CHANGE 1982–1988	% CHANGE 1982–1988
CHALK	245	216	–29	–12
CHINA CLAY	51	53	2	4
CLAY/SHALE	758	496	–262	–35
OPENCAST COAL	179	163	–16	–9
GYPSUM/ANHYDRITE	13	22	9	69
IGNEOUS ROCK	210	126	–84	–40
IRONSTONE	130	97	–33	–25
LIMESTONE/DOLOMITE	615	521	–94	–15
OIL GAS EXPLORATION	–	78	78	–
OIL/GAS PRODUCTION	–	79	79	–
SAND & GRAVEL	2369	1533	–836	–35
INDUSTRIAL SAND	123	87	–36	–29
SANDSTONE	407	351	–56	–14
SLATE	25	36	11	44
VEIN MINERALS	278	124	–154	–55
OTHER MINERALS	330	361	31	9
TOTAL	5733	4343	–1390	–24

12.6 The number of clay/shale workings decreased by 262 sites (a 35% reduction). This was largely due to reductions in the South East region (101 sites) and in the East Midlands region (55 sites). Vein minerals sites decreased by 154 (55% reduction), largely because of the closure of many small sites in the Peak District National Park. Limestone quarries decreased by 94 sites (a 15% reduction).

Surface Disposal of Mineral Working Deposits

12.7 There were 474 sites permitted for the surface disposal of spoil from mineral workings in 1988, 38% fewer than the 770 sites recorded in the 1982 survey.

12.8 Sites for tipping of wastes from deep mined coal were the most numerous, accounting for 54% of the total (255 sites). Limestone accounted for 11% (54 sites), china clay 10% (46 sites), and igneous rock 7% (34 sites) as illustrated in Figure 12.2 and Table 12.3.

12.9 All mineral types had fewer disposal sites than in 1982 with the exception of opencast coal and "other minerals". These increases were however insignificant (4 in total).

12.10 Vein mineral sites decreased by the largest number, 75 sites, 77% less than in 1982 (Table 12.3). This was due to a reduction in the very large number of small sites within the Peak District National Park. Igneous rock and deep mined coal sites decreased by 65 and 56 sites respectively.

12.11 The effect of these reductions has been to change the relative importance of the major minerals as illustrated in Figure 12.2. Deep mined coal was even more dominant in 1988 than in 1982, with an increase from 41% to 54% of the total. Igneous rock and vein mineral sites in contrast decreased in importance since 1982 from 13% each, to 7% and 5% respectively.

FIGURE 12.2 NUMBER OF SITES FOR THE SURFACE DISPOSAL OF MINERAL WORKING DEPOSITS IN ENGLAND IN 1982 AND 1988 – MAIN MINERAL TYPES.

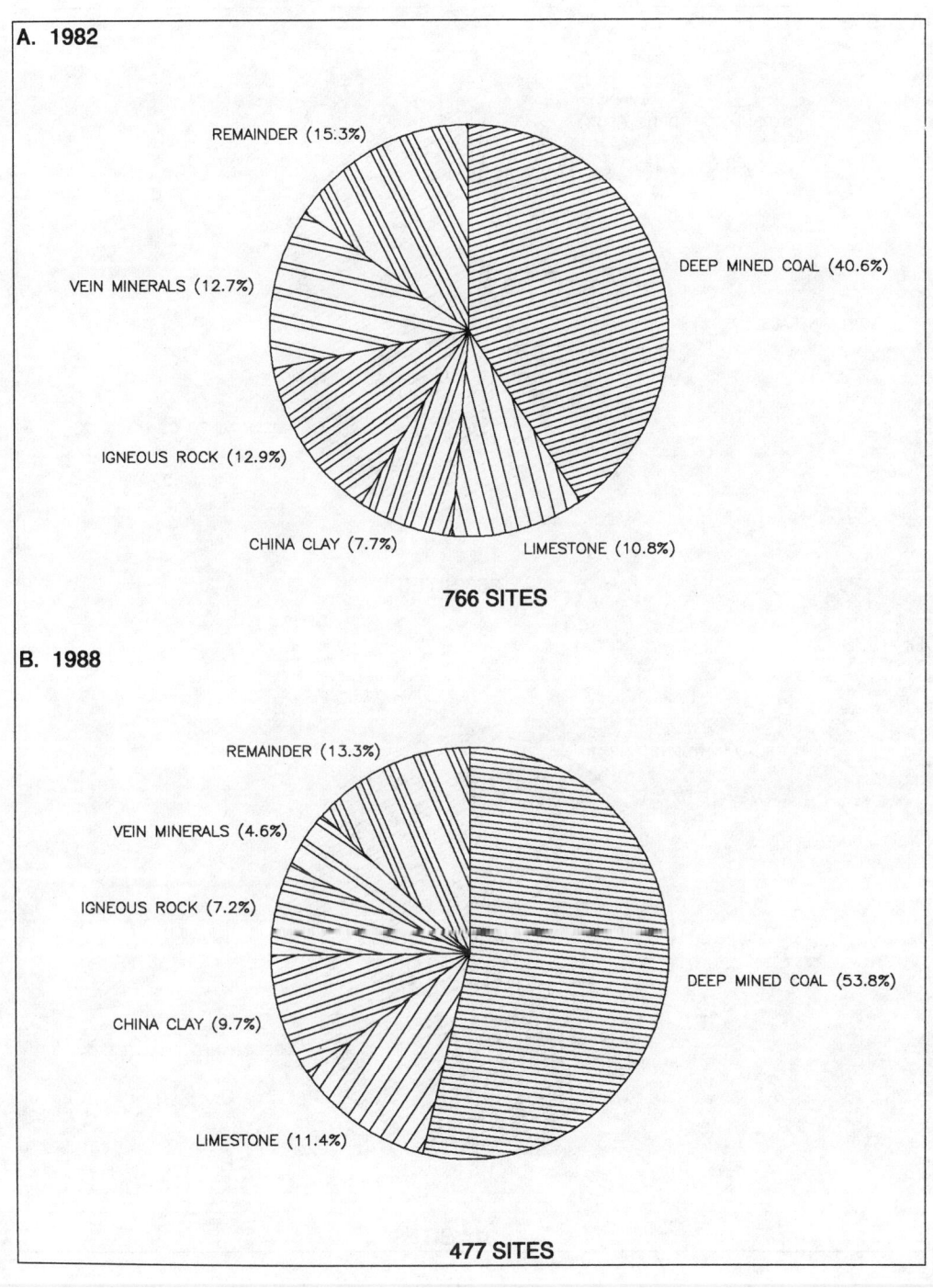

FIGURE 12.3 NUMBER OF SITES FOR UNDERGROUND MINING IN ENGLAND IN 1982 AND 1988 – MAIN MINERAL TYPES

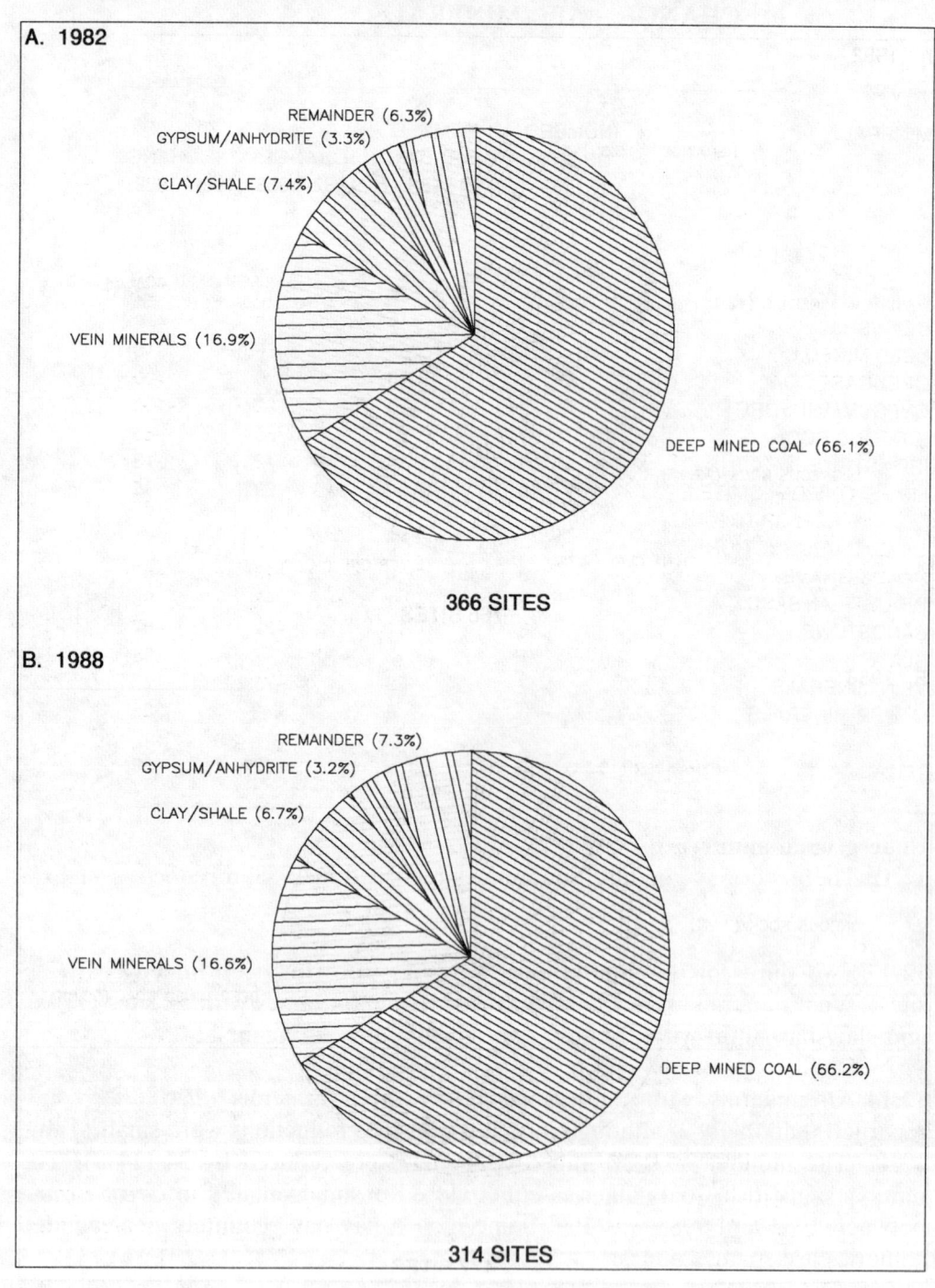

TABLE 12.3 NUMBER OF SITES FOR THE SURFACE DISPOSAL OF MINERAL WORKING DEPOSITS IN ENGLAND IN 1982 AND 1988 INDICATING CHANGE IN NUMBER AND PERCENTAGE CHANGE – ALL MINERALS.

MINERAL	NUMBER OF SITES IN 1982	NUMBER OF SITES IN 1988	CHANGE 1982–1988	% CHANGE 1982–1988
CHINA CLAY	59	46	–13	–22
CLAY/SHALE	13	7	–6	–46
DEEP MINED COAL	311	255	–56	–18
OPENCAST COAL	3	6	3	100
GYPSUM/ANHYDRITE	2	0	–2	–100
IGNEOUS ROCK	99	34	–65	–66
IRONSTONE	16	14	–2	–13
LIMESTONE/DOLOMITE	83	54	–29	–35
OIL GAS EXPLORATION	–	0	0	–
OIL/GAS PRODUCTION	–	0	0	–
SAND & GRAVEL	18	6	–12	–67
INDUSTRIAL SAND	8	1	–7	–88
SANDSTONE	33	11	–22	–67
SLATE	20	13	–7	–35
VEIN MINERALS	97	22	–75	–77
OTHER MINERALS	4	5	1	25
TOTAL	766	474	–292	–38

Underground mining

12.12 There were 314 underground mines recorded in 1988, a reduction of 14% since 1982.

12.13 Two thirds of all underground workings were for deep mined coal (208 sites). Vein minerals were the second most common mines with 52 sites (17%), and clay/shale third with 21 sites (7%) as illustrated in Figure 12.3.

12.14 All minerals with the exception of "other minerals" had fewer sites recorded than in 1982, although in many cases the reductions were small (Table 12.4). The number of deep mined coal sites were reduced by 34 (14%). The number of British Coal collieries either closed or amalgamated in *Great Britain* between 1982 and 1988 was 106. Some mpa's may have counted amalgamated collieries as separate mines.

Regional comparisons

12.15 The largest number of sites associated with mineral workings (surface workings, spoil disposal areas and underground mines) were within the South West region (a total of 986 sites), as illustrated in Figure 12.4.

12.16 The South East region had the largest number of surface mineral workings (926 sites), and East Anglia the smallest with only 276 sites.

12.17 The East Midlands and Yorkshire and Humberside recorded the largest number of underground mines (81 and 79 respectively), whilst no mines were recorded in East Anglia and only 6 in the South East.

TABLE 12.4 NUMBER OF UNDERGROUND WORKINGS IN ENGLAND IN 1982 AND 1988 INDICATING CHANGE IN NUMBER AND PERCENTAGE CHANGE – ALL MINERALS.

MINERAL	NUMBER OF SITES IN 1982	NUMBER OF SITES IN 1988	CHANGE 1982–1988	% CHANGE 1982–1988
CHALK	1	0	–1	–100
CLAY/SHALE	27	21	–6	–22
DEEP MINED COAL	242	208	–34	–14
GYPSUM/ANHYDRITE	12	10	–2	–17
IRONSTONE	5	3	–2	–40
LIMESTONE/DOLOMITE	7	5	–2	–29
VEIN MINERALS	62	52	–10	–16
OTHER MINERALS	10	15	5	50
TOTAL	366	314	–52	–14

FIGURE 12.4 NUMBER OF SURFACE MINERAL WORKINGS, AREAS FOR THE SURFACE DISPOSAL OF MINERAL WORKING DEPOSITS AND UNDERGROUND MINES BY STANDARD REGION IN ENGLAND IN 1988.

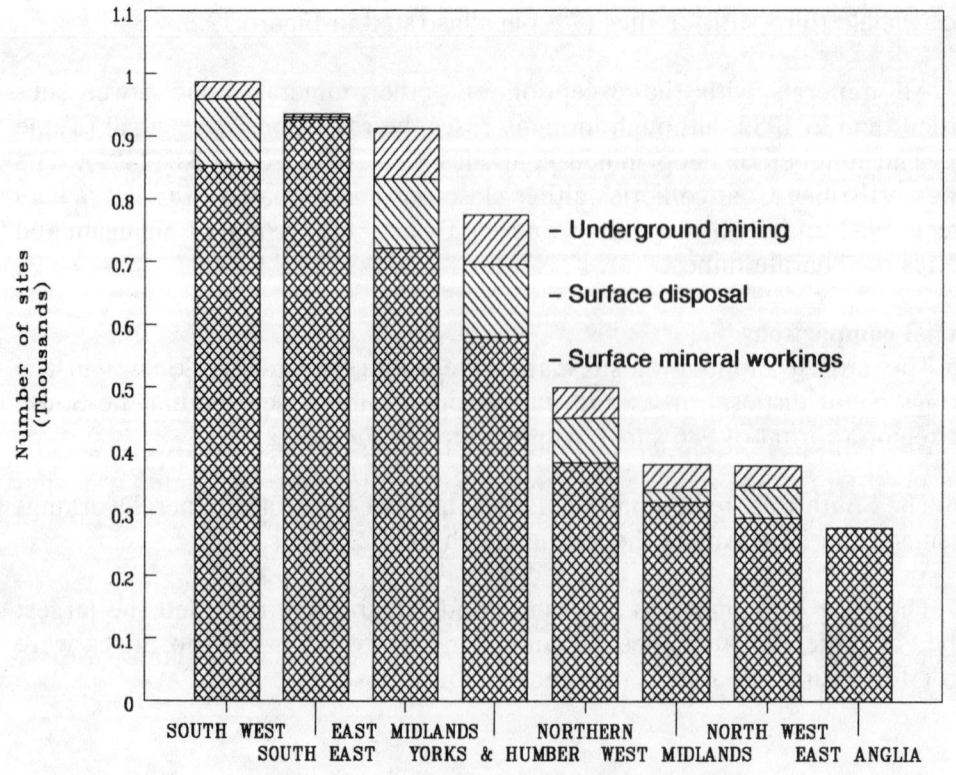

13. REMOVAL OF MATERIAL FROM MINERAL WORKING DEPOSITS

[Information on the reworking of mineral deposits was recorded on Table 3 of the survey form and has been collated into Tables J1 to J8 in volumes 2 and 3 of the survey results].

13.1 The reworking of mineral working deposits (ie former waste tips which still contain worthwhile amounts of saleable mineral) was brought under planning control by Section 1 of the 1981 Minerals Act. Prior to implementation of Section 1 of the Act in May 1986, such activities did not require planning permission. Where operators wished to continue reworking deposits other than stockpiles after that date, applications for planning permission had to be lodged with the mineral planning authorities by November 1986. Sites worked under Class XXVII of the GDO at the base date of this survey (1 April 1988) should only be those awaiting a planning decision (Class A), or under Class C, the reworking of small (less than 2 ha) or temporary (no material deposited more than 5 years previously) mineral working deposits.

13.2 A total of 83 sites were recorded, involving an area of 1270 ha. Of this 55 sites (1040 ha) were being worked under specific planning permissions, the remaining 28 sites (230 ha) where still being worked under the GDO (Table 13.1).

TABLE 13.1 NUMBER OF SITES AND TOTAL AREA PERMITTED FOR THE REMOVAL OF MATERIAL FROM MINERAL WORKING DEPOSITS.

MINERAL	MINERAL PLANNING PERMISSIONS		GDO CLASS XXVII	
	NUMBER OF SITES	AREA (ha)	NUMBER OF SITES	AREA (ha)
CHALK	0	0	1	4
CLAY/SHALE	1	11	0	0
DEEP MINED COAL	33	487	9	213
LIMESTONE	5	47	0	0
SAND & GRAVEL	2	13	0	0
SANDSTONE	4	27	0	0
VEIN MINERALS	7	437	3	5
OTHER MINERALS	3	14	15	7
TOTAL	55	1036	28	229

Note: The mineral type recorded reflects the one involved in the original activity which gave rise to the deposit.

13.3 Deep mined coal tips accounted for half of the sites (42 sites) and 51% of the total area (700 ha), as shown in Figure 13.1. Nine of these sites were worked under the GDO. Vein mineral workings, reworked for fluorspar and to a smaller extent metals such as tin, lead and zinc, involved 10 sites and covered an area of 440 ha.

13.4 The largest number of sites still being worked under the GDO were classified under "other minerals". These were, however, small in area, with 15 sites covering a total area of only 7 ha.

FIGURE 13.1 NUMBER AND AREA OF MINERAL WORKING DEPOSITS PERMITTED IN ENGLAND IN 1988 INDICATING WHETHER OPERATIONS ARE UNDER SPECIFIC PLANNING PERMISSION OR GDO SITES. – MAIN MINERAL TYPES.

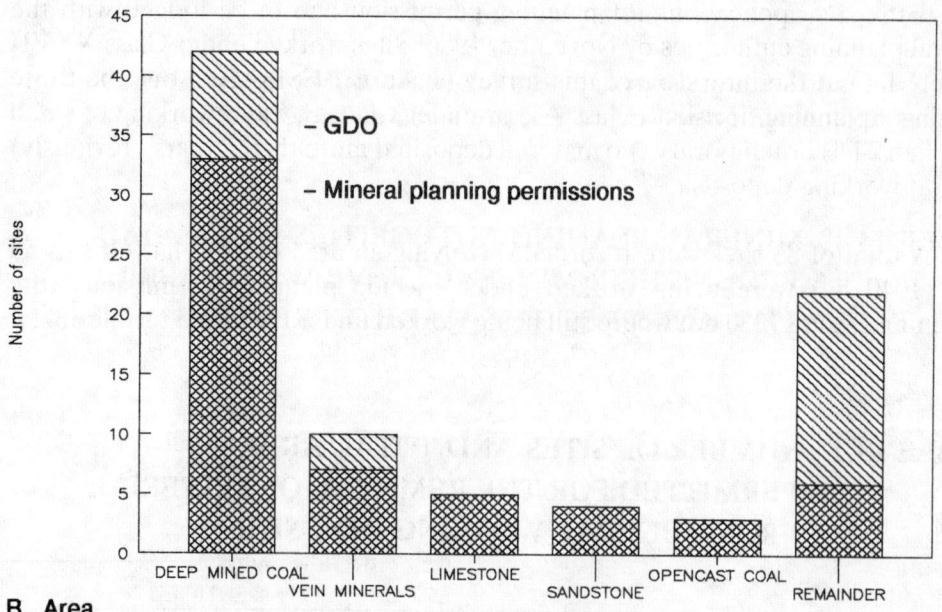

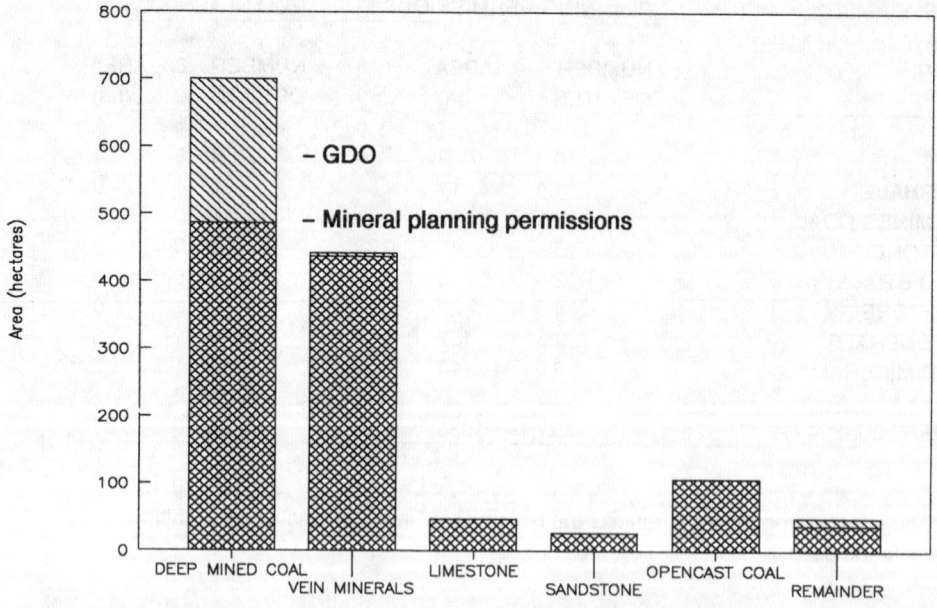

14. REVIEWS UNDER THE TOWN AND COUNTRY PLANNING ACT 1971

[Information on mineral reviews were recorded on Table 6 of the survey form and have been collated in Tables K1 to K8 in Volumes 2 and 3 of the survey results].

14.1 The 1981 Minerals Act inserted section 264A into the 1971 Town and Country Planning Act (section 105 of the 1990 Act). This imposed a duty on mpa's to undertake periodic reviews of mineral working sites in their area and to amend as they saw fit existing planning permissions to raise the environmental standards of the workings (through the mechanism of formal orders or using voluntary agreements, or if appropriate, new planning permissions). These new powers and duties of mpa's came into force in May 1986.

14.2 There is no statutory start date for the reviews and by the 1st April 1988 only 15 mpa's had begun the process. These are listed in Table 14.1.

TABLE 14.1 MINERAL PLANNING AUTHORITIES WHICH HAD BEGUN THEIR MINERALS REVIEW BY 1 APRIL 1988, AND PROGRESS MADE ON REVIEWS.

MINERAL PLANNING AUTHORITY	NUMBER OF SITES/ PERMISSIONS REVIEWED	NUMBER OF SITES/ PERMISSIONS IDENTIFIED FOR FURTHER ACTION	NUMBER OF SITES/ PERMISSIONS WHERE TYPE OF ACTION IDENTIFIED
BARNSLEY MBC	0	0	0
CORNWALL	11	0	0
CUMBRIA	0	0	0
DERBYSHIRE	103	0	0
DEVON	67	0	0
HAMPSHIRE	62	49	0
HUMBERSIDE	184	58	4
ISLE OF WIGHT	43	0	0
KENT	108	23	0
LAKE DISTRICT NATIONAL PARK	16	8	0
LEICESTERSHIRE	15	12	2
NORFOLK	125	9	2
SUFFOLK	0	0	1
SURREY	57	0	0
WAKEFIELD MBC	27	16	0
TOTAL	807	175	9

14.3 The reviews should include unimplemented permissions and all sites which have been active within five years of the start date of the review. There is no set order for sites to be reviewed and it is not known whether the sites most likely to require further action were selected first. If the "worst" sites have been reviewed first, the percentage of sites identified for further action may be disproportionately high when compared with all the sites in the authorities' areas, as many sites may not have been reviewed by the survey's base date.

14.4 At the time of the survey date 750 sites and 68 unimplemented permissions had been reviewed. Of these 171 sites (23%) and 4 unimplemented permissions (6%) had been identified for further action. Only 9 sites or unimplemented permissions had had specific forms of action identified, and in all cases this involved voluntary agreements. No orders were reported as having been made.

14.5 Table 14.2 lists the number of sites and unimplemented permissions for each mineral type which had been reviewed, indicating the number identified for further action. The relative proportion of each mineral type reviewed will depend to a large extent on the particular mpa's which made returns. In view of the relatively small number of mpa's which returned information on mineral reviews it is inappropriate to draw many conclusions at this stage.

TABLE 14.2 NUMBER OF SITES AND UNIMPLEMENTED PERMISSIONS REVIEWED BY MINERAL PLANNING AUTHORITIES BY 1 APRIL 1988, AND PROGRESS MADE ON REVIEWS.

MINERAL	NUMBER OF SITES/ PERMISSIONS REVIEWED	NUMBER OF SITES/ PERMISSIONS IDENTIFIED FOR FURTHER ACTION	NUMBER OF SITES/ PERMISSIONS WHERE TYPE OF ACTION IDENTIFIED
CHALK	76	34	0
CHINA CLAY	11	0	0
CLAY/SHALE	84	23	1
DEEP MINED COAL	18	0	0
OPENCAST COAL	31	0	0
IGNEOUS ROCK	9	3	1
IRONSTONE	9	8	0
LIMESTONE/DOLOMITE	53	12	0
OIL/GAS EXPLORATION	26	0	0
OIL/GAS PRODUCTION	1	0	0
SAND & GRAVEL	373	76	6
INDUSTRIAL SAND	50	4	11
SANDSTONE	38	8	0
SLATE	12	5	0
VEIN MINERALS	5	0	0
OTHER MINERALS	22	2	0
TOTAL	818	175	19

14.6 Almost 46% of all sites/permissions reviewed were for sand and gravel working (373), 10% were clay/shale (84) and 9% were chalk workings (76).

14.7 Of the 76 chalk workings reviewed, 45% (34 sites) had been identified for further action. Equivalent figures for clay/shale and sand and gravel were 27% and 20% respectively. Ironstone had the highest proportion requiring further action, 89%, however this related to the review of only 9 sites.

14.8 More recent figures from a survey conducted by the County Planning Officers' Society indicate that a total of only 35 mpa's had begun their reviews by December 1989 (ie a further 20 had started in the succeeding 21 months from April 1988). As part of the commitment to examine the operation of the provisions of the 1981 Mineral Act within 5 years of full implementation, the Department will be evaluating the concerns about lack of progress on the statutory reviews of mineral working sites, the complexity of the order making powers and the minerals compensation provisions.

PART 2 REGIONAL COMMENTARY

15. NORTHERN REGION

Introduction

15.1 The Northern region is made up of 10 mineral planning authorities; four shire counties (Cleveland, Cumbria, Durham and Northumberland); five metropolitan district councils within Tyne and Wear (Gateshead, Newcastle City, North Tyneside, South Tyneside and Sunderland) and the Lake District Special Planning Board (a planning authority in its own right). Mineral workings were recorded by all mpa's.

15.2 Mineral workings were recorded in a number of National Parks and AONB's within the region: the Lake District National Park; the Northumberland National Park and Northumberland Coast AONB in Northumberland; the North Yorkshire Moors National Park in Cleveland, and the Solway Coast and Arnside and Silverdale AONB's in Cumbria.

Surface Mineral Workings

15.3 The Northern region had the fifth largest area of permissions for surface mineral workings in England, covering an area of 11280 ha and involving 380 sites. This was 12% of the total permitted area for surface mineral workings in England.

15.4 There were 11 mineral categories recorded in the region (including "other minerals"). Permissions for opencast coal sites covered the largest area (4570 ha) accounting for 41% of the total permitted area of surface mineral workings in the region (Figure 15.1). Limestone was the second most widespread mineral with permissions covering 2440 ha (22% of the total permitted area) whilst "other minerals" (1260 ha) and sand and gravel (1060 ha) permissions accounted for 11% and 9% respectively.

15.5 Over half (54%) of the total area in England permitted for opencast coal workings, and 21% of limestone permissions were in the Northern region (Table 15.1).

TABLE 15.1 TOTAL PERMITTED AREA FOR SURFACE MINERAL WORKINGS IN THE NORTHERN REGION INDICATING THE PERCENTAGE REGIONAL CONTRIBUTION TO THE ENGLAND TOTAL.

MINERAL	TOTAL AREA PERMITTED IN REGION (ha)	TOTAL AREA PERMITTED IN ENGLAND (ha)	REGIONAL CONTRIBUTION TO ENGLAND TOTAL (%)
CLAY/SHALE	590	10090	5.8
OPENCAST COAL	4570	8420	54.3
GYPSUM/ANHYDRITE	20	810	2.5
IGNEOUS ROCK	610	2240	27.2
LIMESTONE	2440	11490	21.2
SAND & GRAVEL	1060	29040	3.7
INDUSTRIAL SAND	190	2130	8.9
SANDSTONE	190	2940	6.5
SLATE	220	310	71.0
VEIN MINERALS	130	1540	8.4
OTHER MINERALS	1260	6530	19.3
TOTAL	11280		

Note: Areas are rounded to the nearest 10 hectares.

15.6 Permissions for the working of igneous rock and slate covered much smaller areas (610 ha and 220 ha respectively). However, the permitted area for slate accounted for almost three quarters of the England total for this mineral, while permissions for igneous rock were 27% of the England total.

15.7 Over 78% of the permitted area for surface mineral workings had satisfactory reclamation conditions. This was considerably higher than the national average of 63%. With the exception of East Anglia this was the highest percentage of the total permitted area with satisfactory reclamation conditions in England. A further 11% (1250 ha) had reclamation conditions but these were considered to be unsatisfactory. Land with no provision for reclamation covered 960 ha (9%) and land damaged by surface workings and which were unlikely to be reclaimed covered 240 ha (2%) (Figure 15.2).

15.8 Permissions for opencast coal workings were the main reason for the high proportion of the total area with satisfactory reclamation conditions. All 4570 ha permitted had satisfactory reclamation conditions (Figure 15.3).

15.9 Nearly 60% of limestone permissions were considered to be satisfactory (1410 ha). A further 26% (630 ha) had unsatisfactory reclamation conditions while 13% (330 ha) had no provisions for reclamation. Almost all of the slate permissions had no provisions for reclamation (94% or 210 ha).

15.10 There were 470 ha of workings which required imported fill to achieve reclamation, 4% of the permitted area for surface mineral workings in the region. This was low compared to the national average of 10% (Section 7.19). Sand and gravel workings requiring fill covered the largest area (180 ha) and accounted for 39% of the total. Clay/shale and limestone workings both accounted for a further 29% (140 ha each) of the total (Figure 15.4).

FIGURE 15.1 AREA OF SURFACE MINERAL WORKINGS PERMITTED IN THE NORTHERN REGION IN 1988 – MAIN MINERAL TYPES.

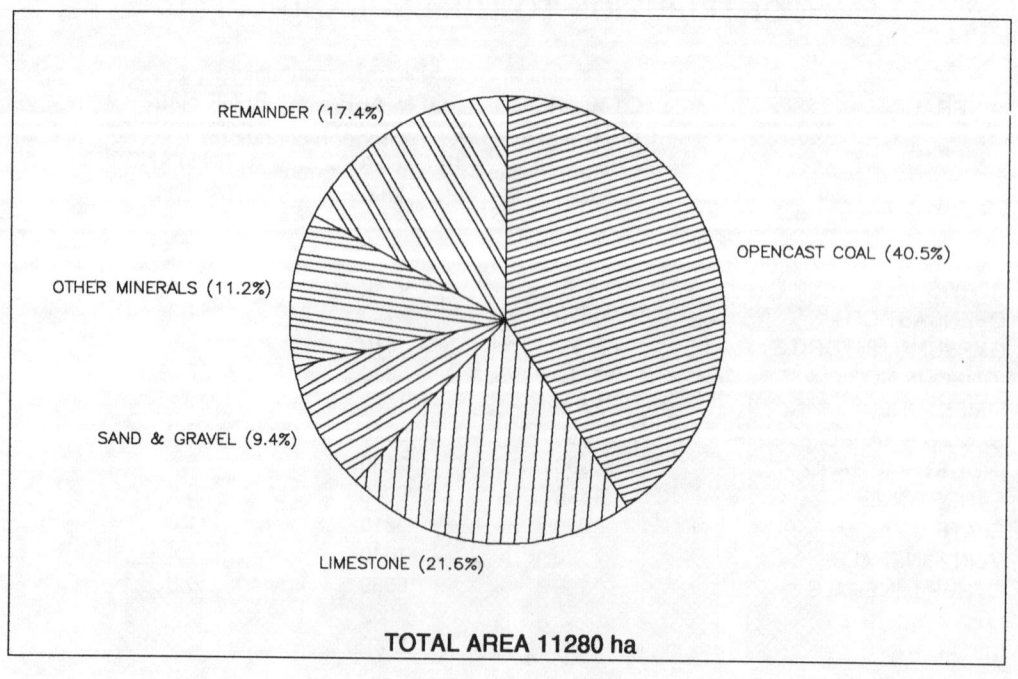

FIGURE 15.2 NATURE OF PROVISIONS FOR RECLAMATION FOR SURFACE MINERAL WORKINGS IN THE NORTHERN REGION IN 1988.

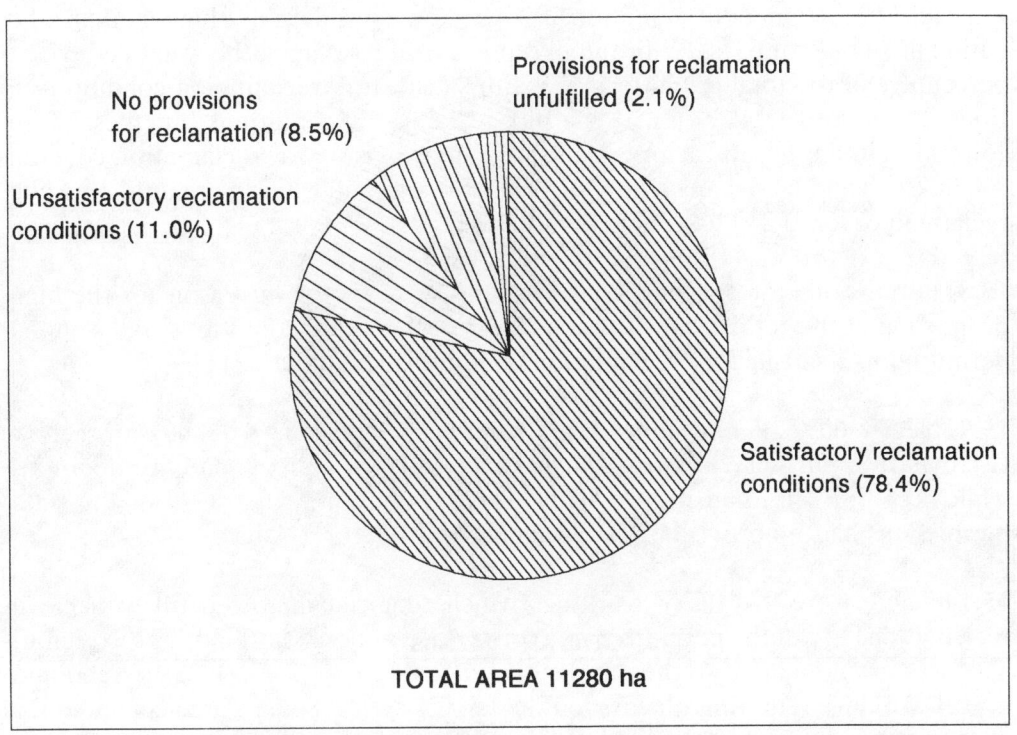

FIGURE 15.3 AREA AND NATURE OF PROVISIONS FOR RECLAMATION OF SURFACE MINERAL WORKINGS IN THE NORTHERN REGION IN 1988 – MAIN MINERAL TYPES.

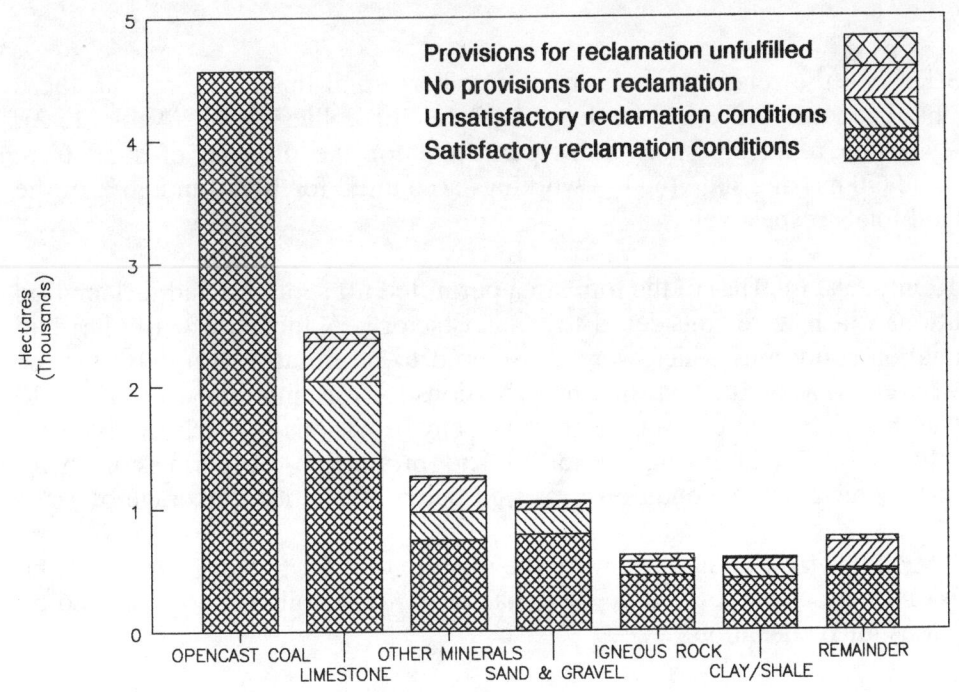

FIGURE 15.4 SURFACE MINERAL WORKINGS DEPENDANT ON IMPORTED FILL IN THE NORTHERN REGION IN 1988.

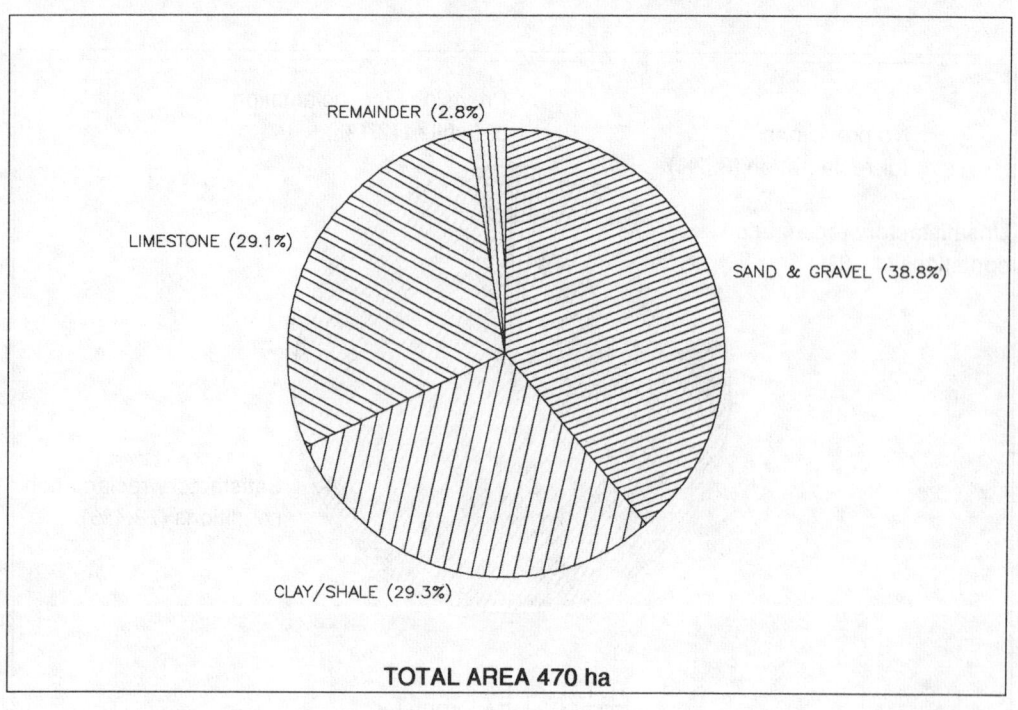

Surface disposal of mineral working deposits

15.11 The Northern region had the fourth largest area permitted for the surface disposal of mineral working deposits (ie spoil tips) in England. Spoil derived from 8 mineral categories were recorded covering a total area of 1570 ha. This was 9% of the total permitted area for spoil tips in England.

15.12 Permissions for the surface disposal of deposits from deep mined coal accounted for 87% (1370 ha) of the total area permitted in the region (Figure 15.5).

15.13 Despite the relatively large area of colliery spoil tips in the region, these accounted for only 12% of the England total for colliery spoil (Table 15.2). However, the relatively small areas permitted for the disposal of spoil from ironstone (70 ha) and slate (60 ha) workings accounted for 100% and 86% of the England totals respectively.

15.14 Only 25% (400 ha) of the total area permitted for spoil tips had reclamation conditions which were considered to be satisfactory. A further 3% (40 ha) had reclamation conditions which were considered to be unsatisfactory. By far the largest area (65% or 1020 ha) had no provisions for reclamation, while 7% (110 ha), had been affected by spoil disposal but any provisions for reclamation were unlikely to be fulfilled (Figure 15.6). The proportion of permissions with satisfactory reclamation conditions was well below the national average of 41%.

15.15 Spoil tips from coal mines were the main reason for the large area with no provisions for reclamation. Almost 880 ha (or 65%) of colliery spoil tips had no provisions for reclamation.

FIGURE 15.5 AREA PERMITTED FOR THE SURFACE DISPOSAL OF MINERAL WORKING DEPOSITS IN THE NORTHERN REGION IN 1988 – MAIN MINERAL TYPES

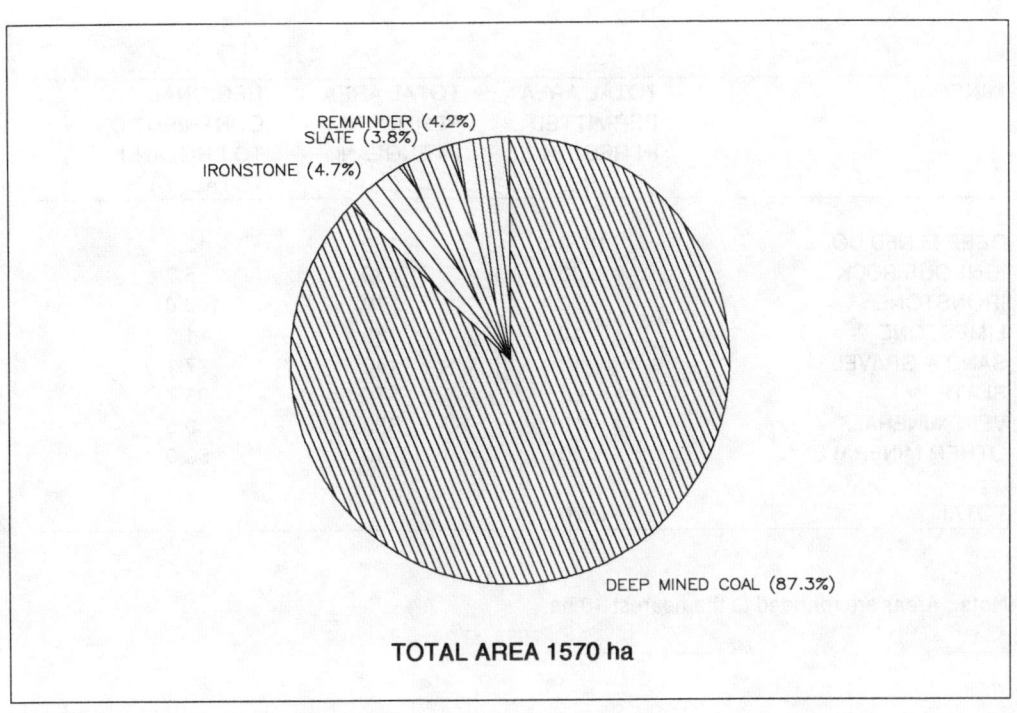

FIGURE 15.6 NATURE OF PROVISIONS FOR RECLAMATION FOR THE SURFACE DISPOSAL OF MINERAL WORKING DEPOSITS IN THE NORTHERN REGION IN 1988.

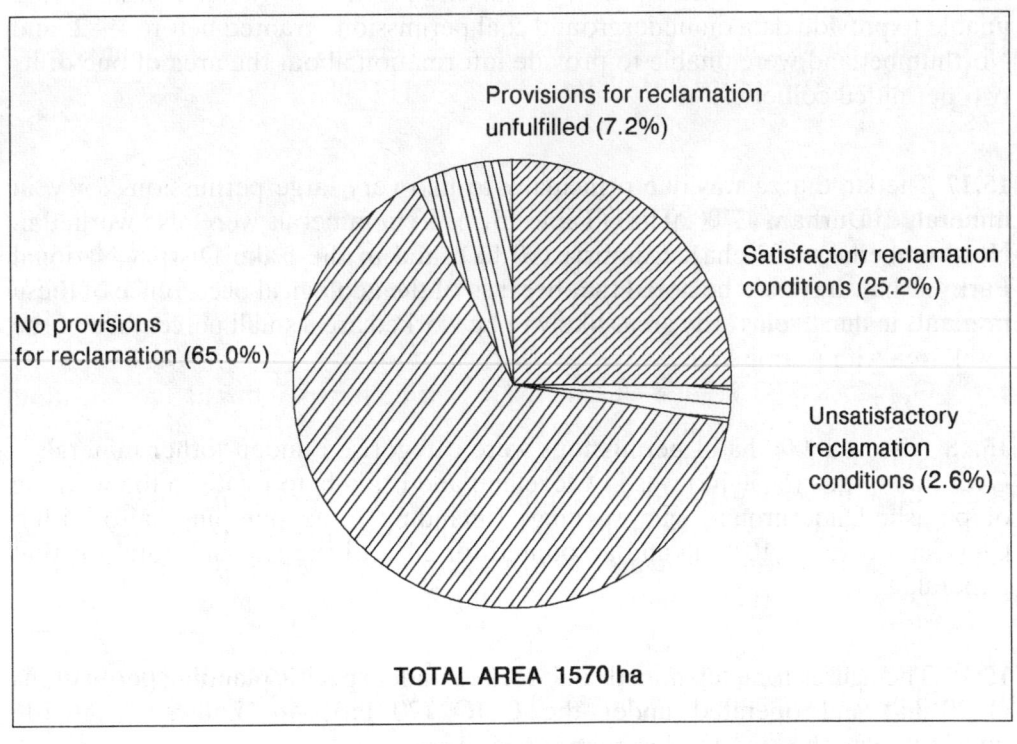

TABLE 15.2 TOTAL PERMITTED AREA FOR THE SURFACE DISPOSAL OF MINERAL WORKING DEPOSITS IN THE NORTHERN REGION, INDICATING THE PERCENTAGE REGIONAL CONTRIBUTION TO THE ENGLAND TOTAL.

MINERAL	TOTAL AREA PERMITTED IN REGION (ha)	TOTAL AREA PERMITTED IN ENGLAND (ha)	REGIONAL CONTRIBUTION TO ENGLAND TOTAL (%)
DEEP MINED COAL	1370	11120	12.3
IGNEOUS ROCK	30	530	5.7
IRONSTONE	70	70	100.0
LIMESTONE	10	780	1.3
SAND & GRAVEL	10	140	7.1
SLATE	60	70	85.7
VEIN MINERALS	<10	320	0.6
OTHER MINERALS	20	40	50.0
TOTAL	1570		

Note: Areas are rounded to the nearest 10 ha.

Underground mining

15.16 The Northern region had by far the largest area of land permitted for underground mining in England, almost 400000 ha. This was despite incomplete information about coal mining in Durham and Northumberland. Durham were unable to provide data on underground coal permissions granted before 1982, and Northumberland were unable to provide information about the area of one of its two permitted collieries.

15.17 The large area was due principally to two very large permissions for vein minerals in Durham (370000 ha) (Table 15.3). Vein minerals were also worked in Northumberland (520 ha), Cumbria (90 ha), and in the Lake District National Park (90 ha). It should be noted that because of the geological occurrence of these minerals in thin "veins" the areas likely to be worked are a small percentage of the total area with permissions.

15.18 Almost 10000 ha of permissions were categorised under "other minerals". Over half of this (5450 ha) was in Cleveland and is likely to relate to the working of potash. Underground salt workings (including brine pumping) also within Cleveland covered 1070 ha and accounted for 24% of the England total for that mineral.

15.19 The figures for deep mined coal worked under specific planning permissions (1320 ha) and operated under the GDO (70 ha), are known to greatly underestimate the true total as mentioned above.

TABLE 15.3 TOTAL AREA OF UNDERGROUND PERMISSIONS IN THE NORTHERN REGION, INDICATING THE PERCENTAGE REGIONAL CONTRIBUTION TO THE ENGLAND TOTAL.

MINERAL	TOTAL AREA PERMITTED IN REGION (ha)	TOTAL AREA PERMITTED IN ENGLAND (ha)	REGIONAL CONTRIBUTION TO ENGLAND TOTAL (%)
COAL (UNDER GDO) (1)	70	315820	0.0
COAL (SPEC PLAN PERM) (2)	1320	46380	2.8
GYPSUM/ANHYDRITE	1340	14670	9.1
SALT	1070	4520	23.7
VEIN MINERAL	369700	379710	97.4
OTHER MINERALS	9960	17850	55.8
TOTAL	383460		

Note: Areas are rounded to the nearest 10 ha.
(1) Excudes Durham and Northumberland.
(2) Excludes Northumberland and permissions in Durham granted before 1982.

Land covered by permissions which include aftercare conditions

15.20 A total of 124 permissions, covering 3490 ha, included aftercare conditions. Surface mineral workings accounted for 118 of the permissions (95%), and covered 99% of this area (3460 ha). Less than 40 ha of land permitted for spoil tips had aftercare conditions (6 permissions). The area of mineral workings covered by aftercare conditions in the region was equivalent to 27% of the region's current permissions. This was high compared with the average for the whole of England of 18%.

15.21 Over three quarters of the area with aftercare conditions (2570 ha) were associated with opencast coal workings (Figure 15.7). Sand and gravel (370 ha) and limestone (250 ha) made up a further 11% and 7% respectively.

15.22 Agriculture was by far the most common intended afteruse for workings covered by aftercare conditions, accounting for 96% of the total area (3340 ha). Amenity afteruses were proposed for 4% of the total area (140 ha) (Figure 15.7). These figures reflect the preference for agricultural reclamation recorded in the region between 1982 and 1988 as discussed in paragraph 15.30. The very high percentage of land proposed for reclamation to agriculture, the highest percentage for any region, compares with the national average of 75% (paragraph 9.13).

Reworking of Mineral Working Deposits

15.23 Six mineral planning authorities within the region had mineral working deposits being reworked in their area. Four mineral types were involved: deep mined coal, limestone, sandstone and "other minerals" (see Table J2 in Volume 2 and Table J4 in Volume 3).

15.24 The total area of 330 ha recorded involved 25 sites. With the exception of two small sites worked under the GDO, all sites were worked under specific planning permissions.

FIGURE 15.7 AREA OF MINERAL WORKINGS COVERED BY AFTERCARE CONDITIONS IN THE NORTHERN REGION IN 1988 – MAIN MINERAL TYPES, AND PROPOSED ENDUSES.

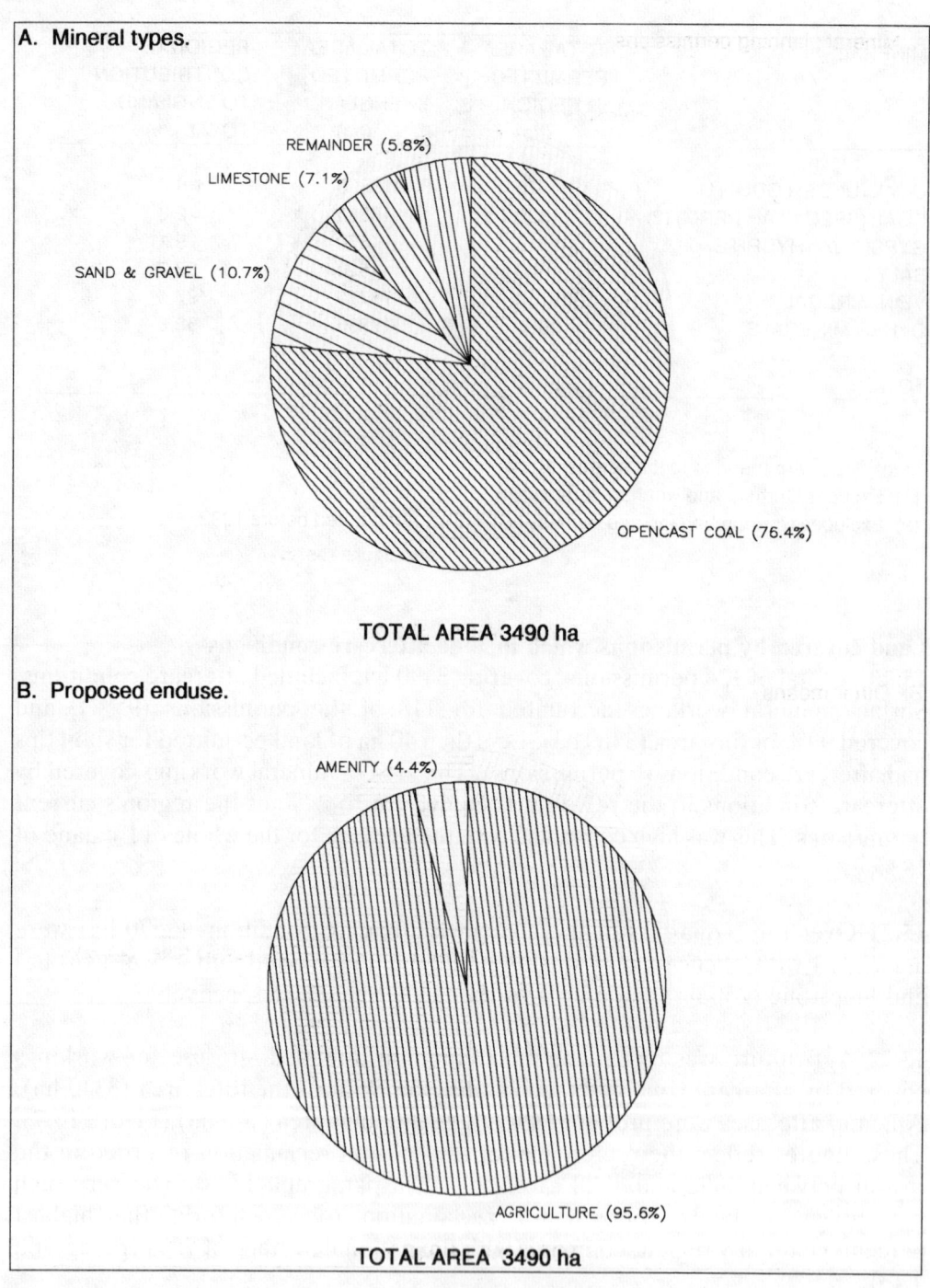

FIGURE 15.8 CHOSEN AFTERUSE OF LAND RECLAIMED BY MINERAL PLANNING PERMISSIONS (A) AND BY OTHER MEANS (B) IN THE NORTHERN REGION BETWEEN 1982 AND 1988.

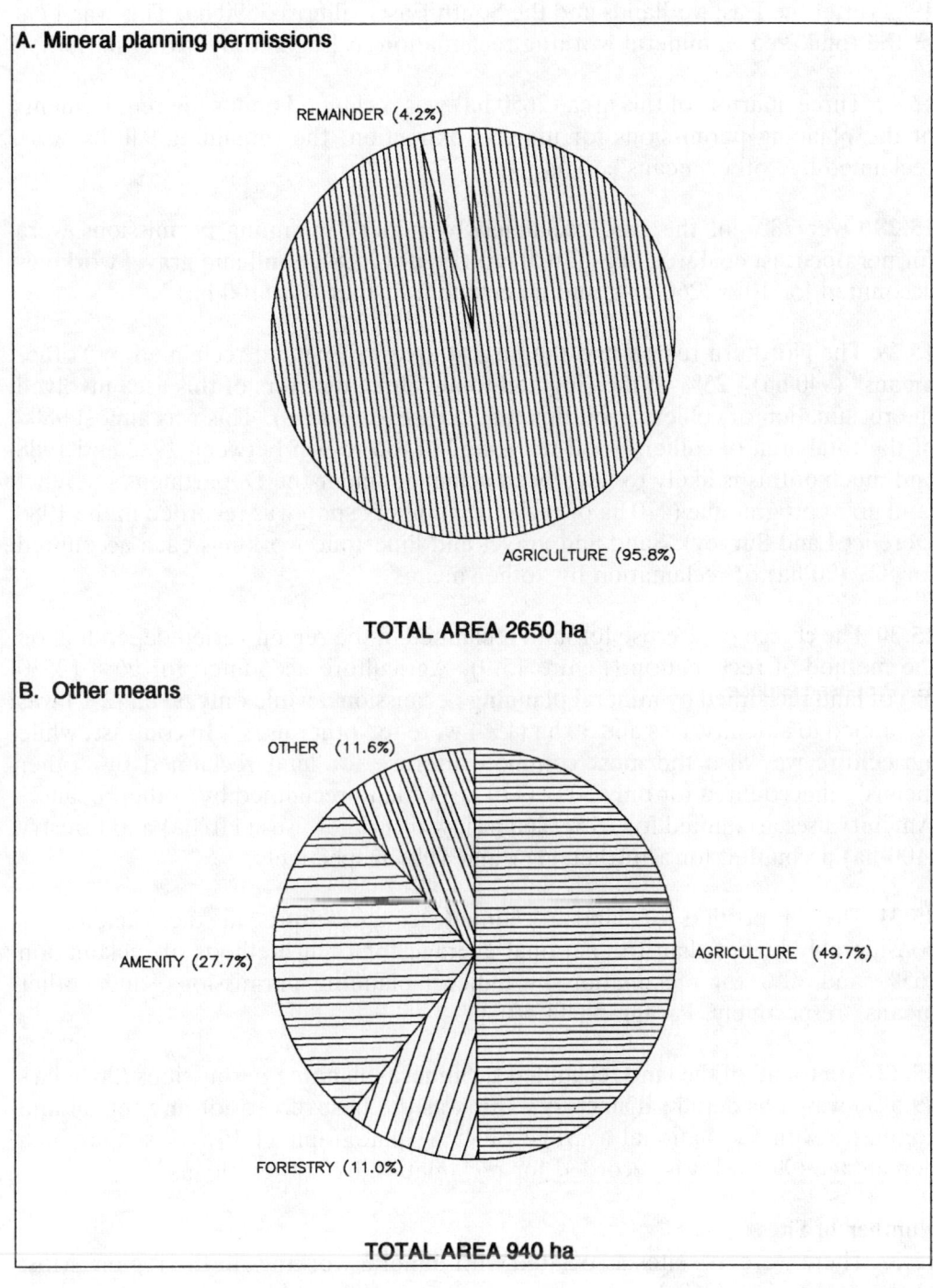

15.25 Eighteen of the sites were reworking deep mined coal spoil tips (250 ha), accounting for over half of the number of such sites and area of reworking of this mineral in England.

Reclamation

15.26 The Northern region had the third largest area in England reclaimed since 1982 (after the East Midlands and the South East), almost 3590 ha. This was 17% of the total area of mineral working reclamation in England over that period.

15.27 Three quarters of this area (2650 ha) was reclaimed under the requirements of the planning permissions for mineral extraction, the remaining 940 ha were reclaimed by "other means".

15.28 Over 78% of the land reclaimed by mineral planning permissions were former opencast coal workings (2070 ha) (Figure 15.8). Sand and gravel workings accounted for 10% (260 ha), and limestone a further 4% (100 ha).

15.29 The Northern region had the largest area in England reclaimed by "other means" (940 ha) - 25% of the England total. Three quarters of this area involved the reclamation of colliery spoil tips (650 ha) (Figure 15.8). This was almost 98% of the total area of colliery spoil reclaimed in the region between 1982 and 1988 and much of this is likely to have been achieved under the Department's derelict land grant programme (410 ha of reclaimed colliery spoil was recorded in the 1988 Derelict Land Survey). Sand and gravel and limestone workings each accounted for 8% (70 ha) of reclamation by "other means".

15.30 The choice of afteruse for land reclaimed in the region varied depending on the method of reclamation (Figure 15.9). Agriculture accounted for 96% (2530 ha) of land reclaimed by mineral planning permissions, while only 60 ha (2%) was reclaimed to amenity uses and 30 ha (1%) were to "other uses". In contrast, while agriculture was also the most common afteruse for land reclaimed by "other means", it accounted for only 50% (470 ha) of land reclaimed by "other means". Amenity uses accounted for 28% (260 ha), whilst other uses (110 ha) and forestry (100 ha) accounted for a further 11% and 10% respectively.

15.31 The proportion of land reclaimed to agriculture in the region was considerably higher than the national average for both methods of reclamation (65% and 30% for reclamation by mineral planning permissions, and "other means" respectively. Paragraph 11.30).

15.32 Almost all of the land reclaimed by mineral planning permissions (2630 ha - 99.5%) was considered satisfactory. This was the best record for any region and compares with the national average of 95% (paragraph 11.19). A similar high percentage (99.6%) was recorded for reclamation by "other means".

Number of Sites

15.33 There were 499 sites associated with mineral workings in the region. Most of these, 380 sites (76%), related to surface mineral workings, 70 were for spoil disposal (14%) and 49 were underground workings (10%).

15.34 The Northern region had one of the smallest number of surface mineral working sites permitted in England (380), only 7% of the total. Most numerous were sand and gravel workings (99), limestone quarries (74), opencast coal workings (67) and igneous rock quarries (32) (Table 15.4).

15.35 There were 70 sites permitted for the surface disposal of mineral working deposits, of which 34 related to deep mined coal (Table 15.4). There were 14 sites permitted for the disposal of spoil from ironstone workings and 10 for slate.

FIGURE 15.9 AREA OF LAND RECLAIMED BY MINERAL PLANNING PERMISSIONS (A) AND BY OTHER MEANS (B) IN THE NORTHERN REGION BETWEEN 1982 AND 1988 – MAIN MINERAL TYPES.

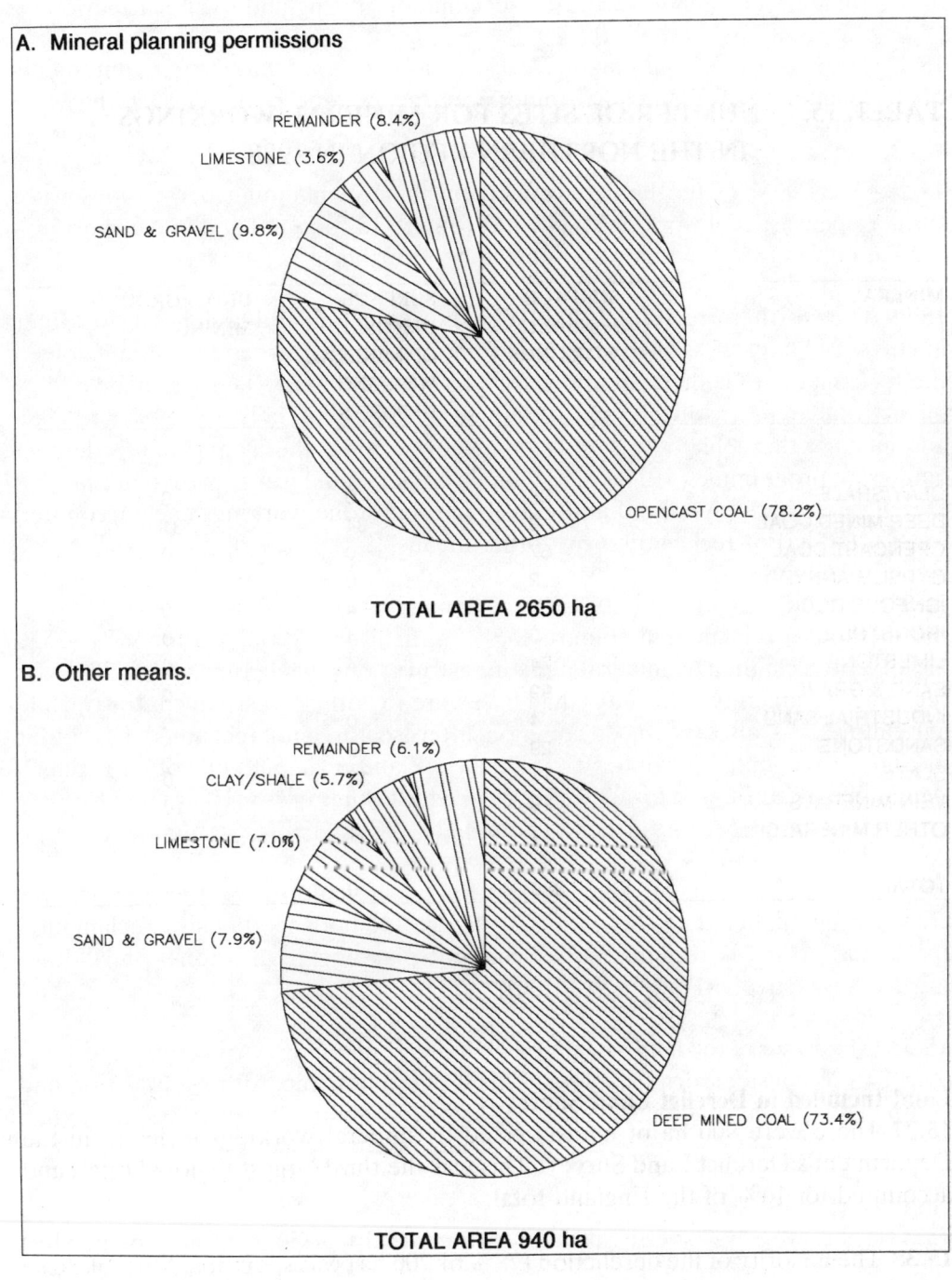

15.36 Forty nine underground workings were recorded involving four mineral types. Deep mined coal was the most common (31 sites). There were 10 vein mineral mines, 5 underground workings for "other minerals" and 3 gypsum/anhydrite mines (Table 15.4).

TABLE 15.4 NUMBER OF SITES FOR MINERAL WORKINGS IN THE NORTHERN REGION IN 1988.

MINERAL	SURFACE MINERAL WORKINGS	SURFACE DISPOSAL OF MINERAL WORKING DEPOSITS	UNDERGROUND MINING
CLAY/SHALE	26	0	0
DEEP MINED COAL	0	34	31
OPENCAST COAL	67	0	0
GYPSUM/ANHYDRITE	2	0	3
IGNEOUS ROCK	32	4	0
IRONSTONE	0	14	0
LIMESTONE	74	4	0
SAND & GRAVEL	99	1	0
INDUSTRIAL SAND	4	0	0
SANDSTONE	29	0	0
SLATE	11	10	0
VEIN MINERALS	19	2	10
OTHER MINERALS	17	1	5
TOTAL	380	70	49

Land Included in Derelict Land Survey

15.37 There were 900 ha of land affected by mineral workings included in the Department's Derelict Land Survey. This was the third largest regional total, and accounted for 16% of the England total.

15.38 The majority of the dereliction (78% or 700 ha) was spoil tips. Most of these (600 ha) had no provisions for reclamation, while a further 100 ha (22%) were sites where provisions for reclamation were unlikely to be fulfilled. There were also 200 ha of surface mineral workings included in the derelict land survey.

15.39 Over two thirds of the total derelict area (610 ha), was spoil disposal from former deep mined coal workings (Figure 15.10). Derelict limestone workings covered 100 ha, of which 40 ha resulted from land with no provisions for reclamation, and 60 ha where the land had been damaged but not reclaimed. Spoil disposal from ironstone workings with no provisions for reclamation contributed almost 70 ha to the region's stock of derelict land.

FIGURE 15.10 LAND DAMAGED BY MINERAL WORKINGS IN THE NORTHERN REGION AND INCLUDED IN THE DERELICT LAND SURVEY IN 1988.

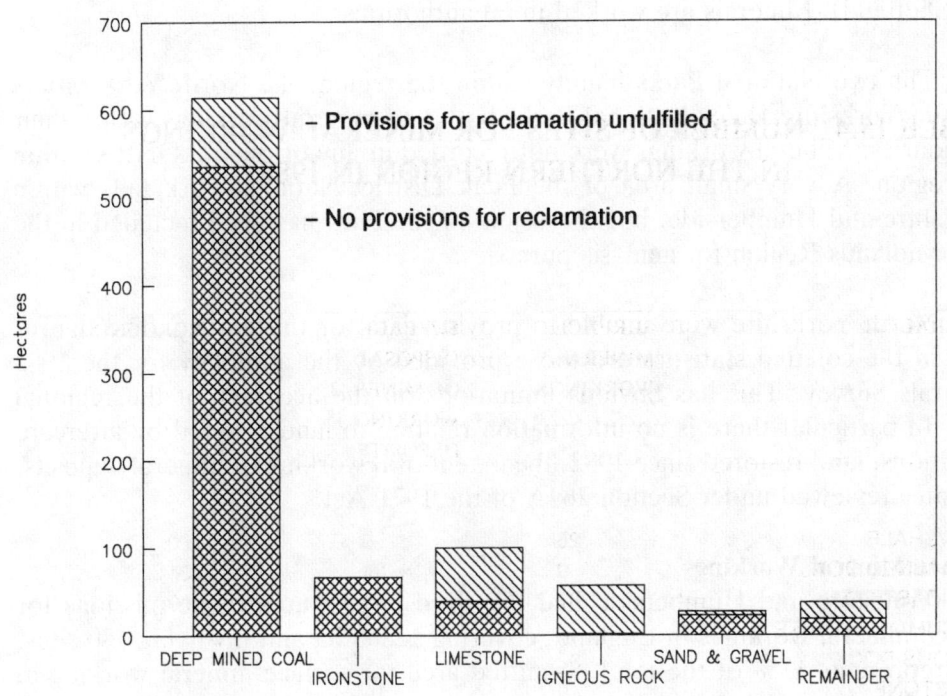

Reviews under Section 264A of the Town and Country Planning Act

15.40 Only two mineral planning authorities had started their review procedures by 1st April 1988, the Lake District Special Planning Board and Cumbria, although Cumbria had not reviewed any sites or permissions by this date.

15.41 Fifteen sites where working had commenced and one unimplemented permission had been reviewed in the Lake District. Eight sites had been identified for further action. No specific action had been determined at the time of the survey.

15.42 Five of the sites where further action was necessary were slate workings. Two igneous rock sites and one sand and gravel working were also identified for further action.

16. YORKSHIRE AND HUMBERSIDE

Introduction

16.1 Yorkshire and Humberside comprises 11 mineral planning authorities; two shire counties (North Yorkshire and Humberside) and nine metropolitan borough councils; five within West Yorkshire (Bradford, Calderdale, Kirklees, Leeds and Wakefield) and four within South Yorkshire (Barnsley, Doncaster, Rotherham and Sheffield). Minerals are worked in all authorities.

16.2 The two National Parks mainly within the region, the North York Moors and the Yorkshire Dales National Parks both had mineral workings within their boundaries. Mineral workings were not recorded in any of the 3 AONB's within the region. A very small area of the Peak District National Park falls within Yorkshire and Humberside, but the whole of the Park has been included in the East Midlands Region for analysis purposes.

16.3 North Yorkshire were unable to provide data for this survey. The figures used in the collated statistics are those provided by the authority for the 1982 Minerals Survey. This has obvious limitations on the accuracy of the regional data. In particular there is no information relating to land covered by aftercare conditions, land restored since 1982, the extent of reworking of mineral deposits, and sites reviewed under Section 264A of the 1971 Act.

Surface Mineral Workings

16.4 Yorkshire and Humberside had the third largest area of permissions for surface mineral workings in England, covering 12200 ha and involving 580 sites. This represents 13% of the total permitted area for surface mineral working in England.

16.5 A wide range of mineral types were recorded involving 13 of the categories used in this survey. Sand and gravel permissions covered the largest area (2790 ha) accounting for 23% of the region's total. "Other minerals" (21% or 2520 ha), limestone (17% or 2080 ha), opencast coal (9% or 1050 ha), ironstone (9% or 1040 ha), clay/shale (8% or 910 ha) and chalk (7% or 810 ha) all covered significant areas (Figure 16.1 and Table 16.1).

16.6 In terms of the contribution to the national total for a particular mineral, "other minerals" were the most important, accounting for 39% of the total area recorded in England for this "remainder" category. Almost all of this related to peat permissions in Doncaster (2270 ha) and Humberside (120 ha). Whilst sandstone permissions were smaller in total area than many of the other mineral types, they accounted for 23% of the England total. Comparable national percentages for chalk and limestone were 22% and 18% respectively (Table 16.1).

16.7 Over 6000 ha of surface mineral workings had permissions which included satisfactory reclamation conditions (49% of the regional total). A further 2110 ha (17%) were covered by reclamation conditions considered to be unsatisfactory. Land with no provisions for reclamation covered 3870 ha (32%) and land damaged by surface workings and which was unlikely to be reclaimed covered 220 ha (2%) (Figure 16.2). With the exception of the South West region, this was the lowest regional percentage of the permitted area for surface mineral workings covered by satisfactory reclamation conditions.

16.8 Limestone, sandstone and sand and gravel workings had the largest areas with unsatisfactory reclamation conditions (Figure 16.3). Almost half of the area of limestone permissions with some form of reclamation conditions was considered to be controlled by unsatisfactory conditions (820 ha). Similarly with

TABLE 16.1 TOTAL AREA PERMITTED FOR SURFACE MINERAL WORKINGS IN YORKSHIRE AND HUMBERSIDE, INDICATING THE PERCENTAGE REGIONAL CONTRIBUTION TO THE ENGLAND TOTAL.

MINERAL	TOTAL AREA PERMITTED IN REGION (ha)	TOTAL AREA PERMITTED IN ENGLAND (ha)	REGIONAL CONTRIBUTION TO ENGLAND TOTAL (%)
CHALK	810	3660	22.1
CLAY/SHALE	910	10090	9.0
OPENCAST COAL	1050	8420	12.5
GYPSUM/ANHYDRITE	70	810	8.6
IGNEOUS ROCK	20	2240	0.9
IRONSTONE	1040	14420	7.2
LIMESTONE	2080	11490	18.1
OIL/GAS EXPLORATION	<10	80	9.2
OIL/GAS PRODUCTION	<10	60	3.3
SAND & GRAVEL	2790	29040	9.6
INDUSTRIAL SAND	230	2130	10.8
SANDSTONE	680	2940	23.1
OTHER MINERALS	2520	6530	38.6
TOTAL	12200		

Note: Areas are rounded to the nearest 10 hectares.

FIGURE 16.1 AREA OF SURFACE MINERAL WORKINGS PERMITTED IN YORKSHIRE AND HUMBERSIDE IN 1988 – MAIN MINERAL TYPES.

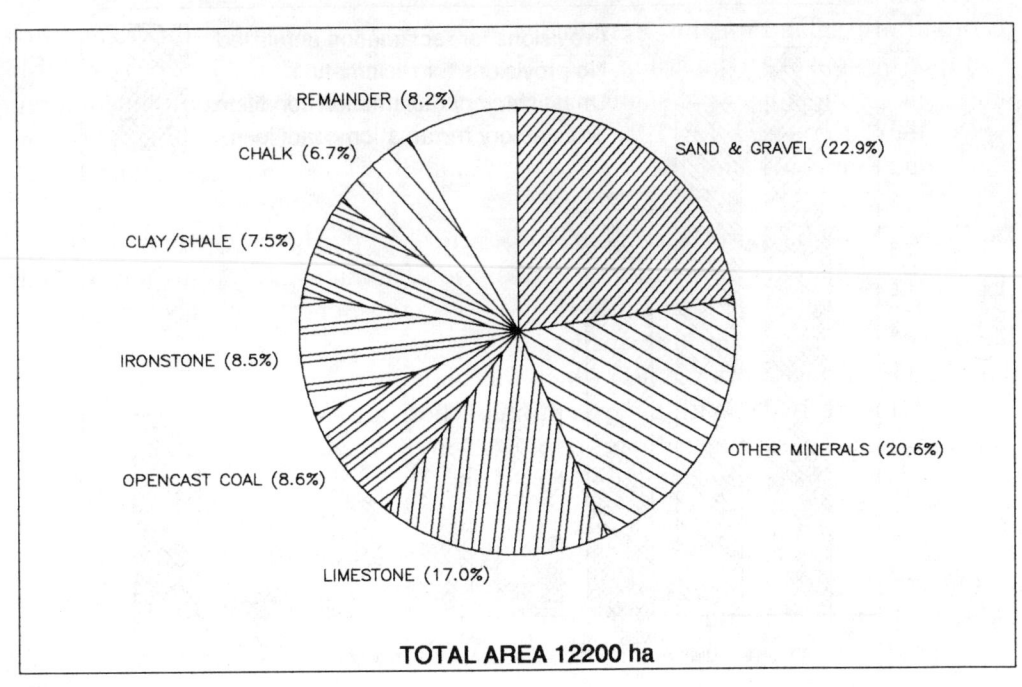

FIGURE 16.2 NATURE OF PROVISIONS FOR RECLAMATION FOR SURFACE MINERAL WORKINGS IN YORKSHIRE AND HUMBERSIDE IN 1988.

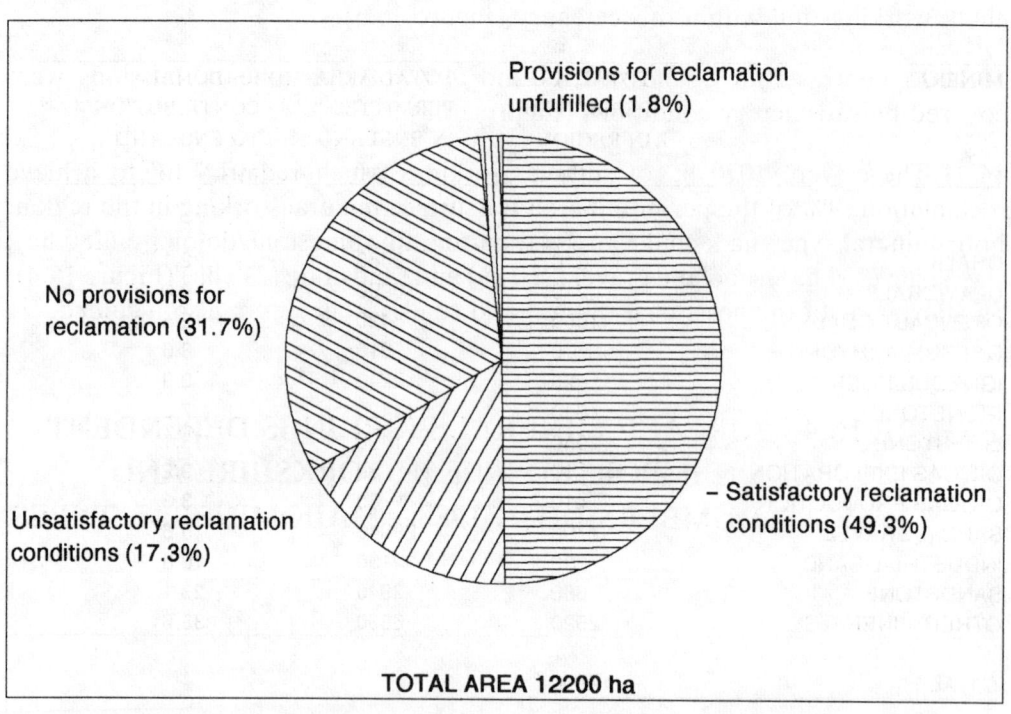

FIGURE 16.3 AREA AND NATURE OF PROVISIONS FOR RECLAMATION OF SURFACE MINERAL WORKINGS IN YORKSHIRE AND HUMBERSIDE IN 1988 –MAIN MINERAL TYPES.

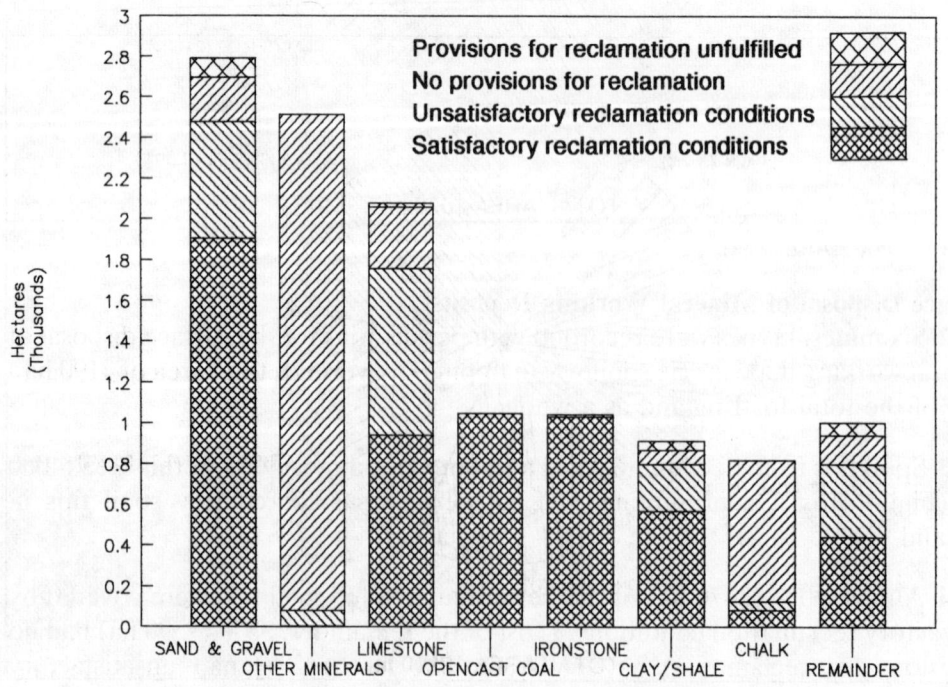

sandstone, over half of the area with reclamation conditions (300 ha) was considered to be of an unsatisfactory standard. Almost a quarter of sand and gravel workings with reclamation conditions (570 ha) were considered unsatisfactory.

16.9 "Other minerals" and chalk had the largest areas with no provisions for reclamation - 2440 ha of "other minerals" (mainly peat workings) and 680 ha of chalk workings fell within this category (Figure 16.3).

16.10 In contrast, all opencast coal and 99% of ironstone permissions were covered by satisfactory conditions (Figure 16.3).

16.11 There were 1070 ha of surface workings which required fill to achieve reclamation, 9% of the permitted area for surface mineral working in the region. Four mineral types had similar areas requiring fill, limestone/dolomite (280 ha), clay/shale (230 ha), sand and gravel (220 ha) and sandstone (210 ha) (Figure 16.4). Smaller areas of opencast coal (80 ha) and chalk (50 ha) were also included.

FIGURE 16.4   SURFACE MINERAL WORKINGS DEPENDENT ON IMPORTED FILL IN YORKSHIRE AND HUMBERSIDE IN 1988 – MAIN MINERAL TYPES.

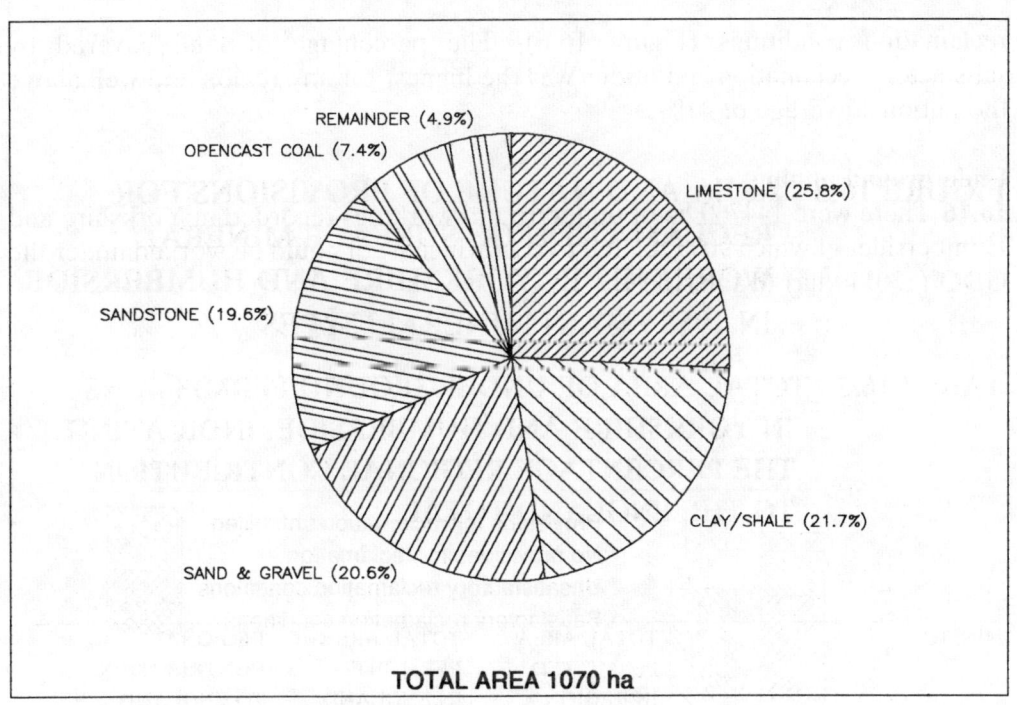

Surface Disposal of Mineral Working Deposits

16.12 Six mineral types were recorded with permissions for the surface disposal of mineral working deposits (ie spoil/waste tipping), covering a total area of 4190 ha - 23% of the total for England as a whole.

16.13 Spoil tips from deep mined coal accounted for almost 97% of this area (4060 ha) (Figure 16.5), representing 37% of permissions for colliery spoil tips in England (Table 16.2).

16.14 Almost 60% (2470 ha) of the permitted area for spoil tips were covered by satisfactory reclamation conditions. Most of the remainder, 38% (1590 ha) had no provisions for reclamation. A small area (90 ha or 2%) had unsatisfactory

TABLE 16.2 TOTAL AREA PERMITTED FOR THE SURFACE DISPOSAL OF MINERAL WORKING DEPOSITS IN YORKSHIRE AND HUMBERSIDE, INDICATING THE PERCENTAGE REGIONAL CONTRIBUTION TO THE ENGLAND TOTAL.

MINERAL	TOTAL AREA PERMITTED IN REGION (ha)	TOTAL AREA PERMITTED IN ENGLAND (ha)	REGIONAL CONTRIBUTION TO ENGLAND TOTAL (%)
CLAY/SHALE	30	350	8.6
DEEP MINED COAL	4060	11120	36.5
OPENCAST COAL	60	130	46.2
LIMESTONE	50	780	6.4
INDUSTRIAL SAND	<10	20	15.0
SANDSTONE	<10	60	1.7
TOTAL	4190		

Note: Areas are rounded to the nearest 10 ha.

reclamation conditions (Figure 16.6). The percentage of land covered by satisfactory reclamation conditions was the highest for any region and well above the national average of 41%.

Underground mining

16.16 There were 144760 ha of underground workings recorded in Yorkshire and Humberside, of which 90% related to coal which is, or could be worked under the GDO (130140 ha) (Table 16.3).

TABLE 16.3 TOTAL AREA OF UNDERGROUND PERMISSIONS IN YORKSHIRE AND HUMBERSIDE, INDICATING THE PERCENTAGE REGIONAL CONTRIBUTION TO THE ENGLAND TOTAL.

MINERAL	TOTAL AREA PERMITTED IN REGION (ha)	TOTAL AREA PERMITTED IN ENGLAND (ha)	REGIONAL CONTRIBUTION TO ENGLAND TOTAL (%)
CLAY/SHALE	560	1470	38.1
COAL (UNDER GDO)	130140	315820	41.2
COAL (SPEC PLAN PERM)	2880	46380	6.2
GYPSUM/ANHYDRITE	2000	14670	13.6
SALT	<10	4520	<0.1
VEIN MINERAL	1400	379710	0.4
OTHER MINERALS	7770	17850	43.5
TOTAL	144760		

Note: Areas are rounded to the nearest 10 ha.

FIGURE 16.5 AREA PERMITTED FOR THE SURFACE DISPOSAL OF MINERAL WORKING DEPOSITS IN YORKSHIRE AND HUMBERSIDE IN 1988. – MAIN MINERAL TYPES.

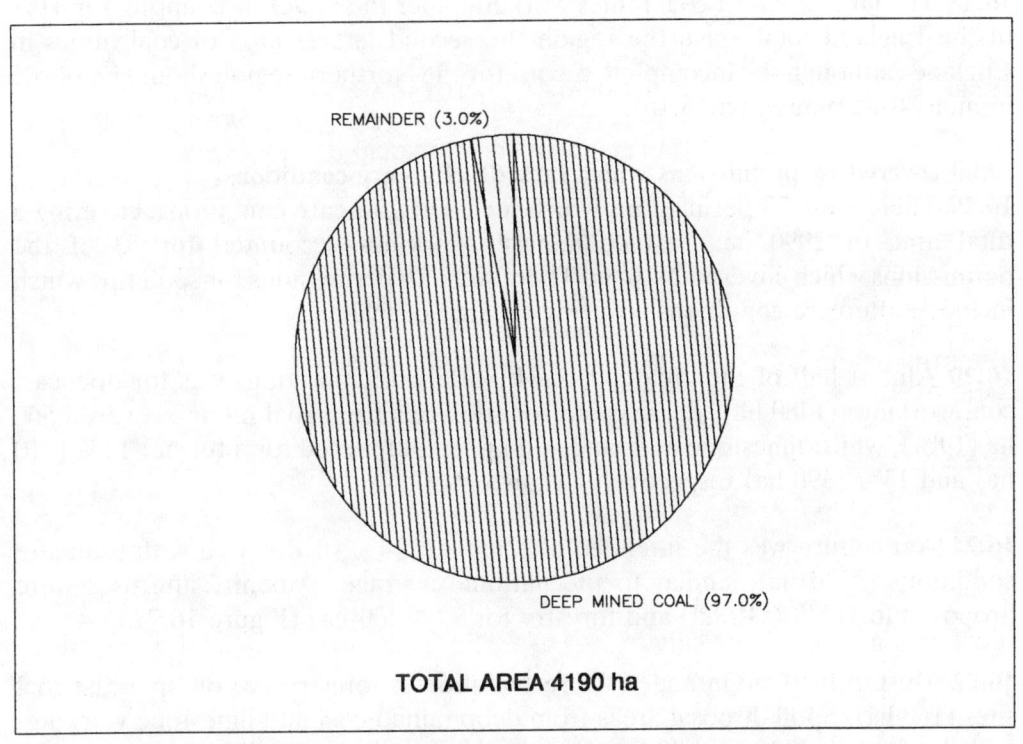

FIGURE 16.6 NATURE OF PROVISIONS FOR RECLAMATION FOR THE SURFACE DISPOSAL OF MINERAL WORKING DEPOSITS IN YORKSHIRE AND HUMBERSIDE IN 1988.

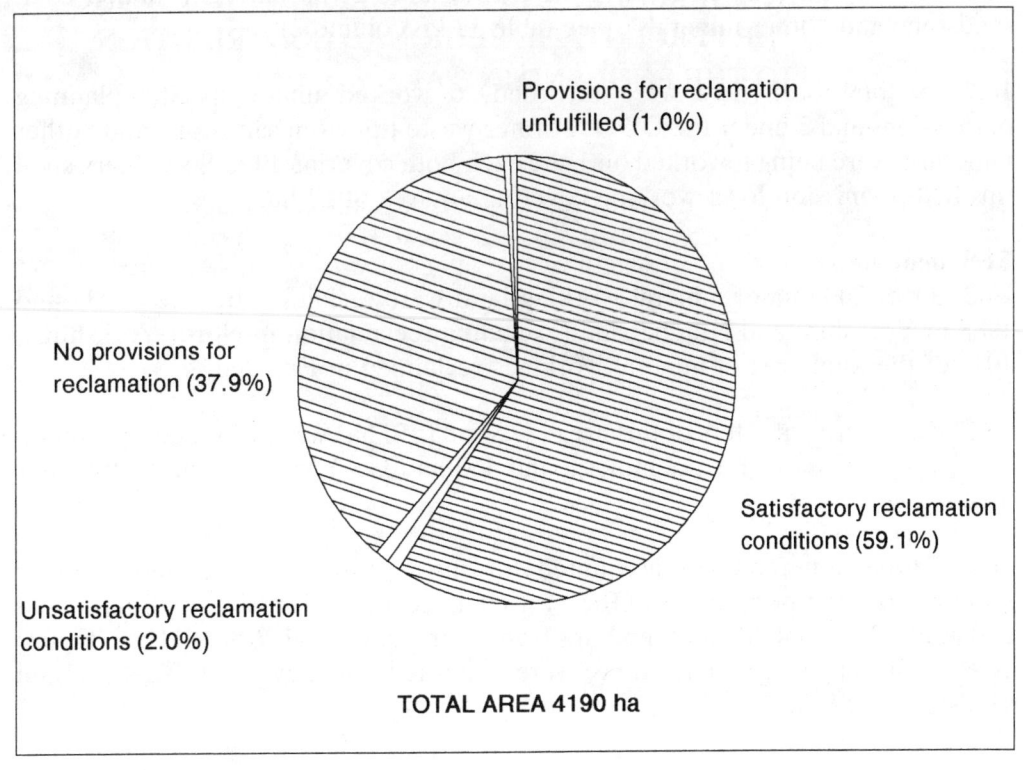

16.17 The second largest mineral category recorded was "other minerals", with permissions covering 7770 ha, 5% of the regions total permitted area. This was all within the North York Moors National Park and is likely to relate to permissions for potash mining. Coal worked under specific planning permissions covered 2880 ha, slightly over 2% of the total area of underground coal permissions.

16.18 The large areas of coal mines worked under the GDO, accounting for 41% of the England total, gave the region the second largest area of coal mines in England (although the incomplete record for the Northern region should be borne in mind - see paragraph 15.16).

Land covered by permissions which include aftercare conditions

16.19 There were 73 permissions which included aftercare conditions, covering a total area of 2990 ha. Surface mineral workings accounted for 53 of the permissions which covered 2410 ha. There were 20 permissions for spoil tips which included aftercare conditions covering an area of 590 ha.

16.20 Almost half of the area covered by aftercare conditions was for opencast coal workings (1480 ha). Permissions for spoil tips from coal mining covered 560 ha (19%), whilst limestone and sand and gravel accounted for a further 14% (410 ha) and 13% (390 ha) respectively (Figure 16.7).

16.21 Agriculture was the intended afteruse for 74% of the area with aftercare conditions (2230 ha), similar to the national average. Amenity afteruses were proposed for 17% (510 ha) and forestry for 9% (260 ha) (Figure 16.7).

16.22 Most of the land intended for reclamation to forestry was on opencast coal sites (160 ha). Spoil disposal areas from deep mined coal and limestone workings added a further 70 ha and 20 ha of forestry respectively. Similarly most amenity afteruses were planned for opencast coal workings (290 ha). In addition 80 ha of limestone, 80 ha of sand and gravel and 50 ha of spoil from deep mined coal were also intended for amenity afteruses.

Reworking of Mineral Working Deposits

16.23 Four mineral planning authorities had mineral working deposits being reworked in their area (Barnsley, Bradford, Leeds and Rotherham) Three mineral types were involved covering a total area of 50 ha; deep mined coal, sandstone and "other minerals" (see table J2 in Volume 2).

16.24 A total of 8 sites were recorded, 6 worked under specific planning permissions and 2 under the GDO. Former waste tips from sandstone and "other minerals" were being reworked on 1 site each both covering 1 ha. Six colliery spoil tips had permission for reworking covering an area of 52 ha.

Reclamation

16.25 Over 2000 ha of mineral working land were reclaimed between 1982 and 1988 in Yorkshire and Humberside (excluding reclamation in North Yorkshire), 10% of the total area of mineral workings reclaimed in England.

16.26 Over three quarters of this area (77% or 1550 ha) was reclaimed by mineral planning permissions for the mineral and 460 ha (33%) was reclaimed by "other means".

16.27 Three mineral types accounted for most of the land reclaimed under mineral planning permissions. Opencast coal accounted for 33% (510 ha), sand and gravel 28% (440 ha) and spoil tips from deep mined coal 26% (400 ha) (Figure 16.8). Smaller but notable areas were recorded for clay/shale (90 ha), and sandstone (60 ha).

FIGURE 16.7 AREA OF MINERAL WORKINGS COVERED BY AFTERCARE CONDITIONS IN YORKSHIRE AND HUMBERSIDE IN 1988 – MAIN MINERAL TYPES, AND PROPOSED ENDUSES.

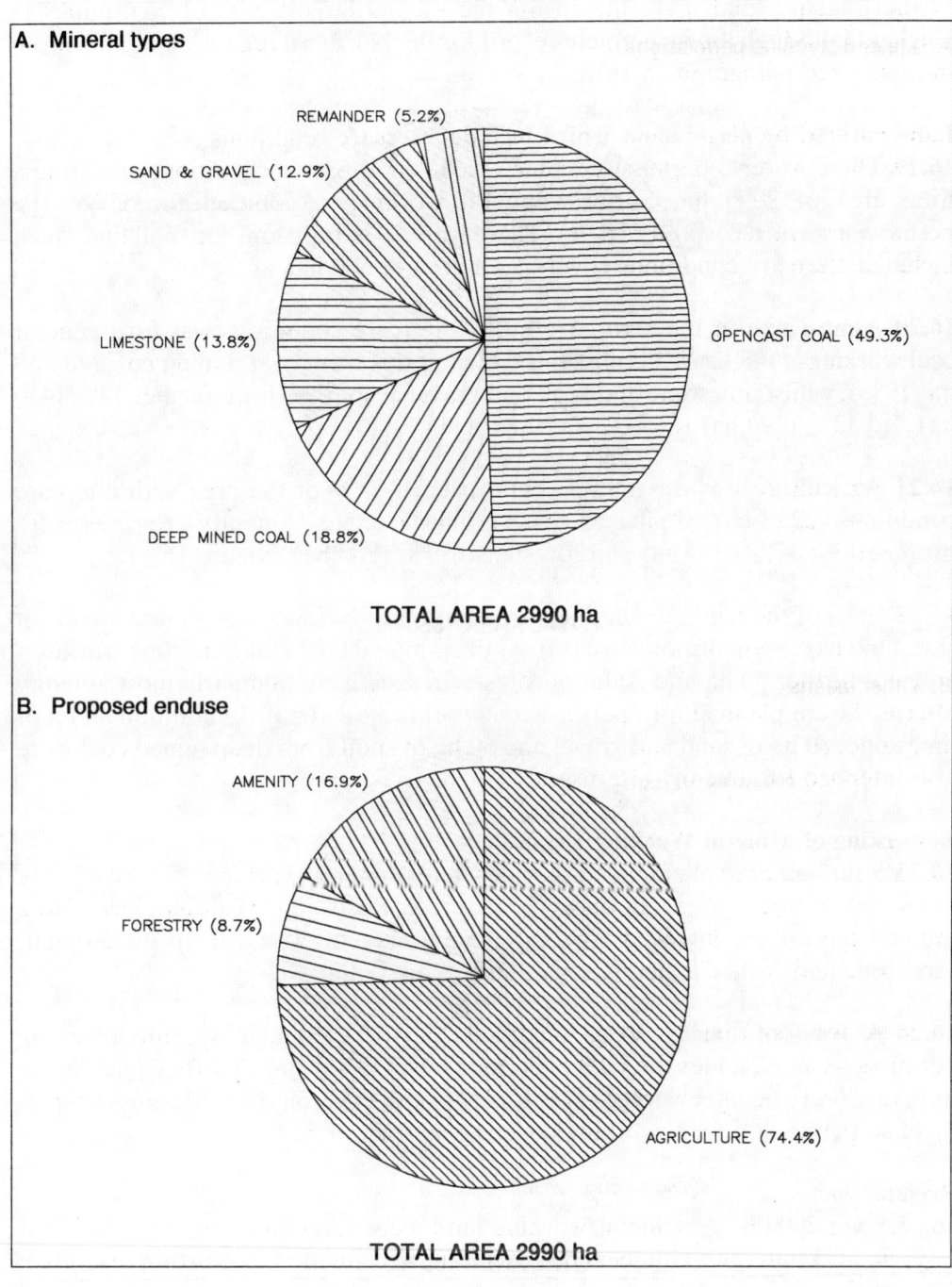

FIGURE 16.8 AREA OF LAND RECLAIMED BY MINERAL PLANNING PERMISSIONS (A) AND BY OTHER MEANS (B) IN YORKSHIRE AND HUMBERSIDE BETWEEN 1982 AND 1988 – MAIN MINERAL TYPES.

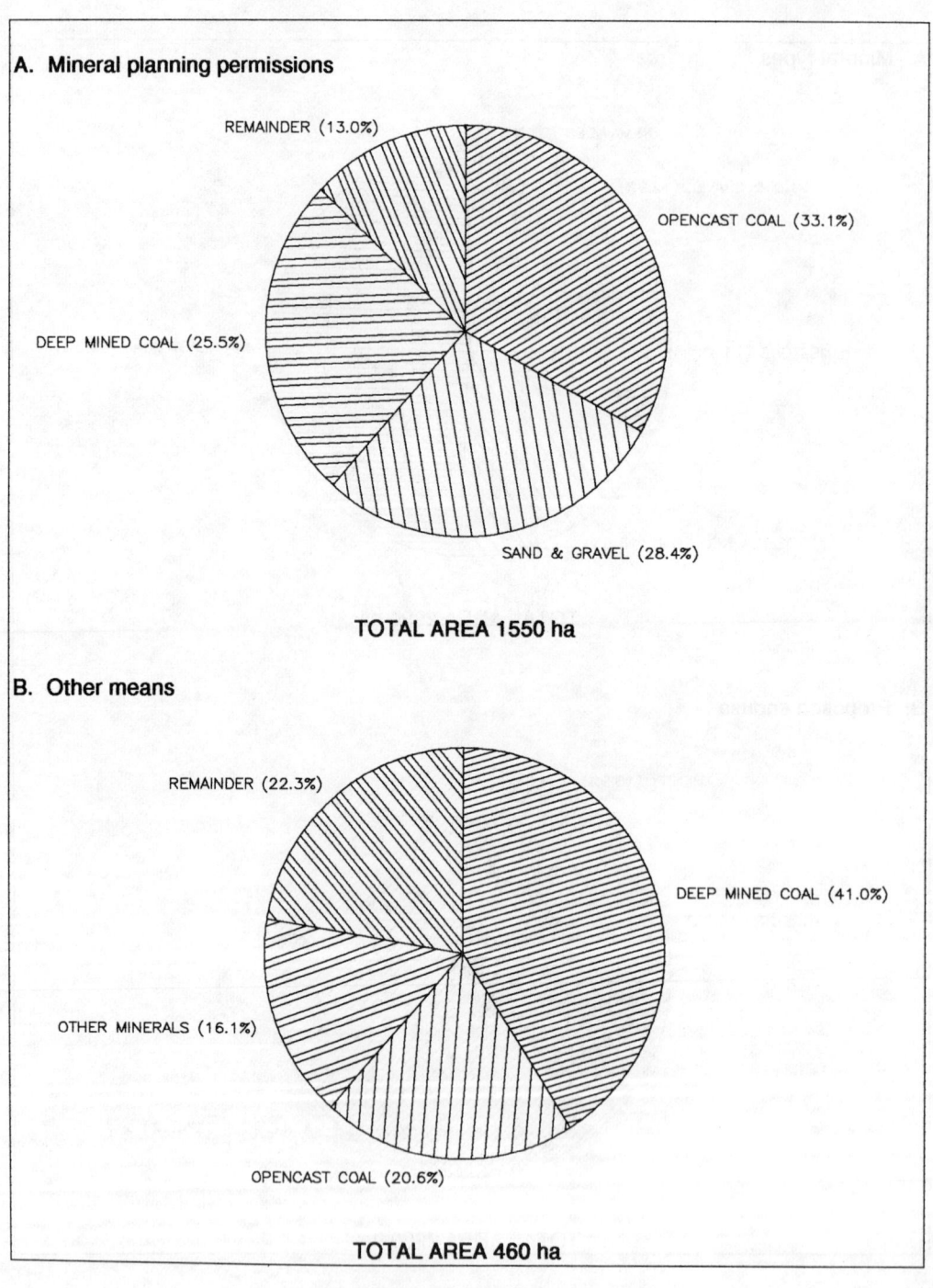

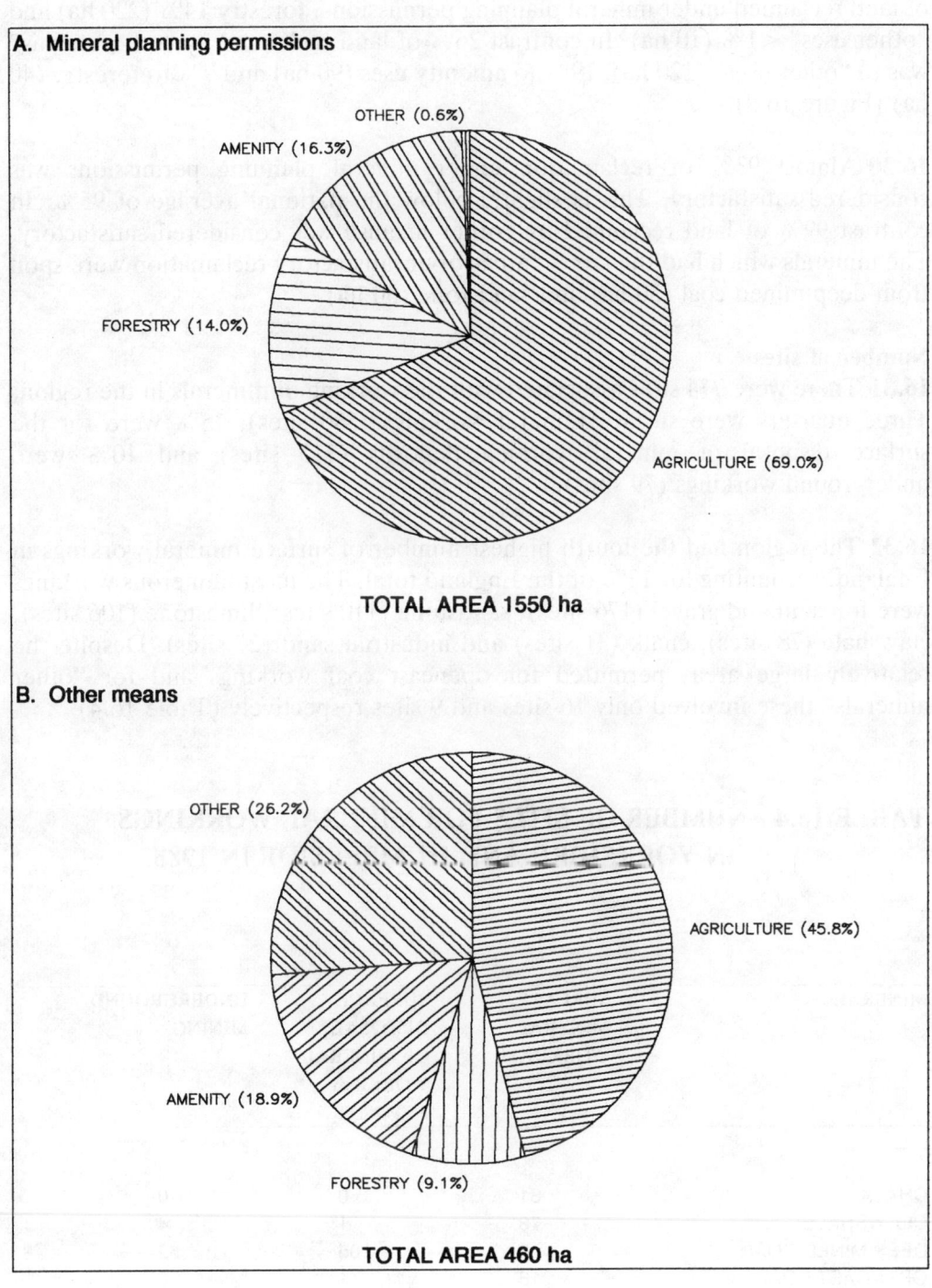

FIGURE 16.9 CHOSEN AFTERUSE OF LAND RECLAIMED BY MINERAL PLANNING PERMISSIONS (A) AND BY OTHER MEANS (B) IN YORKSHIRE AND HUMBERSIDE BETWEEN 1982 AND 1988 – MAIN MINERAL TYPES.

16.28 Forty one percent of mineral land reclaimed by "other means" (190 ha) was spoil from deep mined coal workings. A further 21% (100 ha) were former opencast coal sites and 16% (70 ha) were "other minerals".

16.29 Agriculture was the most common afteruse for land reclaimed by both mineral planning permissions and "other means", accounting for 69% (1070 ha) and 46% (210 ha) respectively. Amenity reclamation accounted for 16% (250 ha) of land reclaimed under mineral planning permissions, forestry 14% (220 ha) and "other uses" <1% (10 ha). In contrast 26% of land reclaimed by "other means" was to "other uses" (120 ha), 19% to amenity uses (90 ha) and 9% to forestry (40 ha) (Figure 16.9).

16.30 Almost 93% of reclamation under mineral planning permissions was considered satisfactory. This is slightly below the national average of 95%. In contrast 99% of land reclaimed by "other means" was considered satisfactory. The minerals which had the largest areas of unsatisfactory reclamation were spoil from deep mined coal (50 ha) and sandstone (30 ha).

Number of sites

16.31 There were 774 sites associated with the working of minerals in the region. Three quarters were surface mineral workings (580 sites), 15% were for the surface disposal of mineral working deposits (115 sites) and 10% were underground workings (79 sites).

16.32 The region had the fourth highest number of surface mineral workings in England, accounting for 13% of the England total. The most numerous workings were for sand and gravel (176 sites), sandstone (110 sites), limestone (106 sites), clay/shale (78 sites), chalk (31 sites) and industrial sand (27 sites). Despite the relatively large areas permitted for opencast coal workings and for "other minerals" these involved only 16 sites and 9 sites respectively (Table 16.4).

TABLE 16.4 NUMBER OF SITES FOR MINERAL WORKINGS IN YORKSHIRE AND HUMBERSIDE IN 1988.

MINERAL	SURFACE MINERAL WORKINGS	SURFACE DISPOSAL OF MINERAL WORKING DEPOSITS	UNDERGROUND MINING
CHALK	31	0	0
CLAY/SHALE	78	1	4
DEEP MINED COAL	0	105	62
OPENCAST COAL	16	1	0
GYPSUM/ANHYDRITE	0	0	1
IRONSTONE	14	0	0
LIMESTONE	106	7	0
OIL/GAS EXPLORATION	12	0	0
OIL/GAS PRODUCTION	1	0	0
SAND & GRAVEL	176	0	0
INDUSTRIAL SAND	27	0	0
SANDSTONE	110	1	0
VEIN MINERALS	0	0	11
OTHER MINERALS	9	0	1
TOTAL	580	115	79

16.33 Over 90% of the 115 sites for the surface disposal of mineral working deposits were related to deep mined coal (105 sites).

16.34 Of the 79 underground workings recorded in the region, 62 were for coal (78%). There were also 11 vein minerals mines, 4 clay/shale, 1 gypsum/anhydrite and one "other minerals" (potash) mine.

Land included in the Derelict Land Survey

16.35 Yorkshire and Humberside had the largest area of derelict land affected by mineral workings recorded in both this and the Department's Derelict Land Survey - almost 1100 ha. This was 25% of the total area of derelict mineral workings recorded in England.

16.36 Spoil tips accounted for 74% of the total area of mineral dereliction in the region (800 ha). Most of this (770 ha), was a result of land being worked with no provisions for reclamation. A further 40 ha had been damaged by spoil disposal where provisions for reclamation had not been carried out. Surface mineral workings added a further 280 ha to the total - 150 ha were derelict because of no provisions for reclamation and 130 ha because provisions for reclamation were unfulfilled. These figures represent 19% of the total area of spoil disposal and 2% of surface mineral workings in the region.

16.37 Spoil derived from deep mined coal workings accounted for 74% of the area recorded in the Derelict Land Survey. Limestone, sand and gravel and sandstone all added a further 70 ha each (Figure 16.10).

FIGURE 16.10 LAND DAMAGED BY MINERAL WORKINGS IN YORKSHIRE AND HUMBERSIDE AND INCLUDED IN THE DERELICT LAND SURVEY IN 1988.

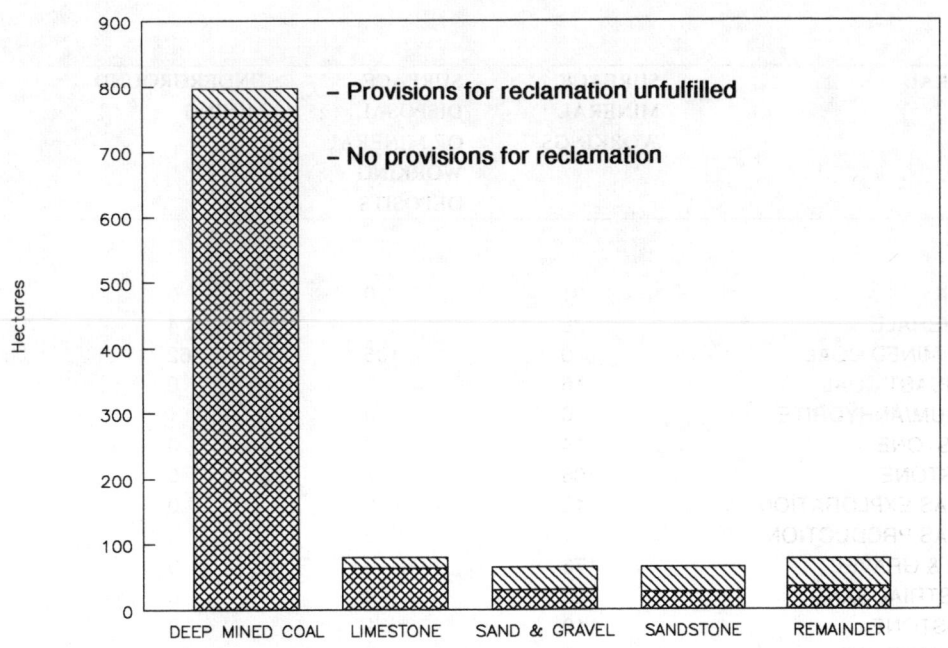

Reviews under Section 264A of the Town and Country Planning Act
16.38 Three mineral planning authorities had started their review procedures by 1st April 1988, Barnsley MBC, Humberside CC, and Wakefield MBC.

16.39 In total 193 sites and 18 unimplemented permissions had been reviewed. Of these, 70 sites (35%) and 4 unimplemented permissions (22%) had been identified for further action. No specific action had been determined at the time of the survey, although 4 voluntary agreements had been made.

16.40 A wide range of the minerals worked in the region had been reviewed, including over 60 sand and gravel sites, 33 chalk and industrial sand workings, and 27 clay/shale sites.

16.41 Over half (54%) of the chalk workings, 37% of clay/shale sites and 35% of sand and gravel sites reviewed had been identified for further action.

17. EAST MIDLANDS REGION

Introduction

17.1 The East Midlands region is made up of six mineral planning authorities; five shire counties (Derbyshire, Leicestershire, Lincolnshire, Northamptonshire and Nottinghamshire), and the Peak District Joint Planning Board (Peak District National Park). Mineral workings were recorded by all six authorities. The Peak District National Park also extends into both the North West region and Yorkshire and Humberside. For the purposes of this analysis all workings within the Park are included in the East Midlands total.

17.2 There was 1 AONB in the region, the Lincolnshire Wolds AONB. Ten surface mineral workings and one underground mine were recorded in the AONB.

Surface Mineral Workings

17.3 The East Midlands had the largest area of surface mineral working permissions in England. Over 26000 ha were recorded, 27% of the total permitted area of surface workings in England.

17.4 A wide range of mineral types are worked within the region, with 14 of the mineral categories used in this survey recorded. Ironstone permissions accounted for almost 50% of this area (12700 ha). Most of the ironstone permissions, are however, unlikely to be worked for the original mineral (see paragraph 7.3). Sand and gravel and limestone permissions covered the second and third largest areas accounting for 21% (5490 ha) and 11% (2920 ha) of the region's total respectively. Opencast coal (6% or 1540 ha), clay/shale (4% or 960 ha) and vein minerals (4% or 930 ha) also covered significant areas (Figure 17.1).

FIGURE 17.1 AREA OF SURFACE MINERAL WORKINGS PERMITTED IN THE EAST MIDLANDS REGION IN 1988 – MAIN MINERAL TYPES.

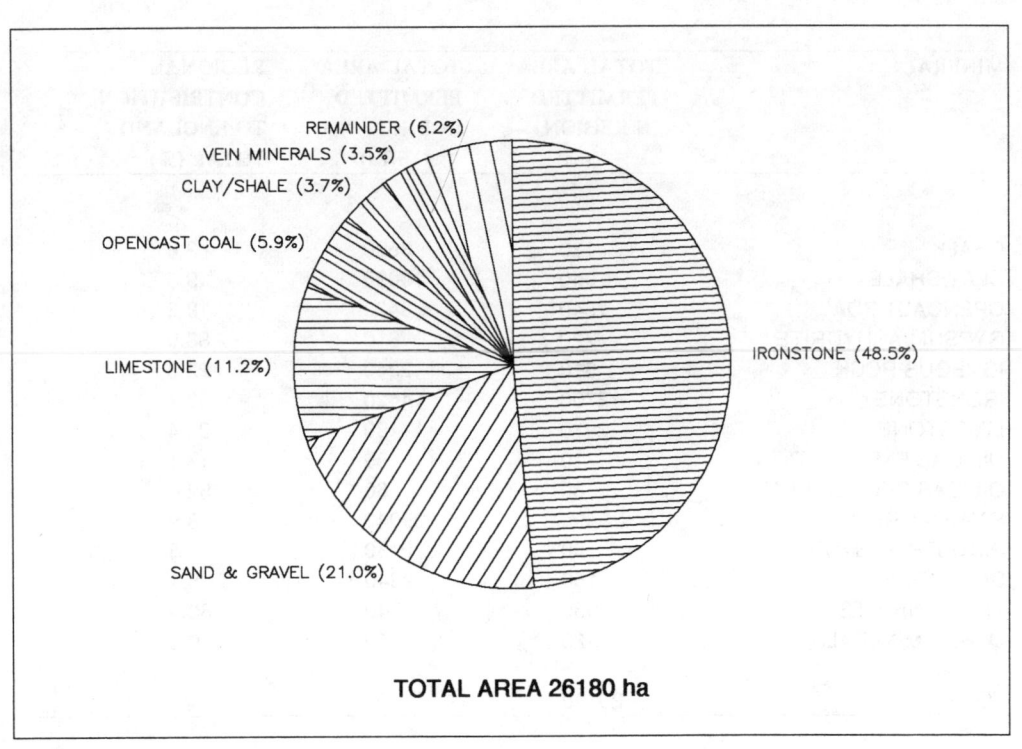

17.5 In terms of the contribution to the national total for individual minerals, the region was important for a number of mineral types. Permissions for gypsum/anhydrite in the region accounted for 89% of the total area permitted for this mineral in England, while ironstone permissions accounted for 88% of the national total. Also significant were surface worked vein minerals (60% of permissions in England), oil/gas production (52%), limestone (25%), igneous rock (22%), sand and gravel (19%) and opencast coal (18%) (Table 17.1).

17.6 Nearly 69% of surface mineral workings had permissions which included satisfactory reclamation conditions (17920 ha), well above the national average of 63%. However a further 5940 ha (23%) had reclamation conditions which were considered to be unsatisfactory. Less than 9% of the permitted area had no provisions for reclamation (2230 ha), and only 0.3% (80 ha) had provisions for reclamation which were unlikely to be fulfilled (Figure 17.2).

17.7 All oil/gas exploration, 98% of opencast coal and 94% of oil/gas production permissions were covered by satisfactory reclamation conditions. Ironstone (76%), sand and gravel (75%), clay/shale (72%) and igneous rock (70%) also had a high proportion of the permitted area with satisfactory reclamation conditions (however, over 3000 ha of ironstone permissions had unsatisfactory reclamation conditions - almost all of which is unworked and is likely to remain so). In contrast, only 30% of limestone permissions, 29% of sandstone, 25% of "other minerals", 24% of chalk, and 12% of vein minerals were considered to have satisfactory reclamation conditions (Figure 17.3).

17.8 The most extensive areas with no provisions for reclamation were permissions for limestone (1250 ha), vein minerals (370 ha), sand and gravel (280 ha), gypsum/anhydrite (110 ha) and clay/shale (90 ha).

TABLE 17.1 TOTAL AREA PERMITTED FOR SURFACE MINERAL WORKINGS IN THE EAST MIDLANDS REGION, INDICATING THE PERCENTAGE REGIONAL CONTRIBUTION TO THE ENGLAND TOTAL.

MINERAL	TOTAL AREA PERMITTED IN REGION (ha)	TOTAL AREA PERMITTED IN ENGLAND (ha)	REGIONAL CONTRIBUTION TO ENGLAND TOTAL (%)
CHALK	130	3660	3.6
CLAY/SHALE	960	10090	9.5
OPENCAST COAL	1540	8420	18.3
GYPSUM/ANHYDRITE	720	810	88.9
IGNEOUS ROCK	510	2240	22.8
IRONSTONE	12700	14420	88.1
LIMESTONE	2920	11490	25.4
OIL/GAS EXPLORATION	10	80	17.1
OIL/GAS PRODUCTION	30	60	52.4
SAND & GRAVEL	5490	29040	18.9
INDUSTRIAL SAND	80	2130	3.8
SANDSTONE	150	2940	5.1
VEIN MINERALS	930	1540	60.4
OTHER MINERALS	20	6530	0.3
TOTAL	26180		

Note: Areas are rounded to the nearest 10 hectares.

FIGURE 17.2 NATURE OF PROVISIONS FOR RECLAMATION FOR SURFACE MINERAL WORKINGS IN THE EAST MIDLANDS REGION IN 1988.

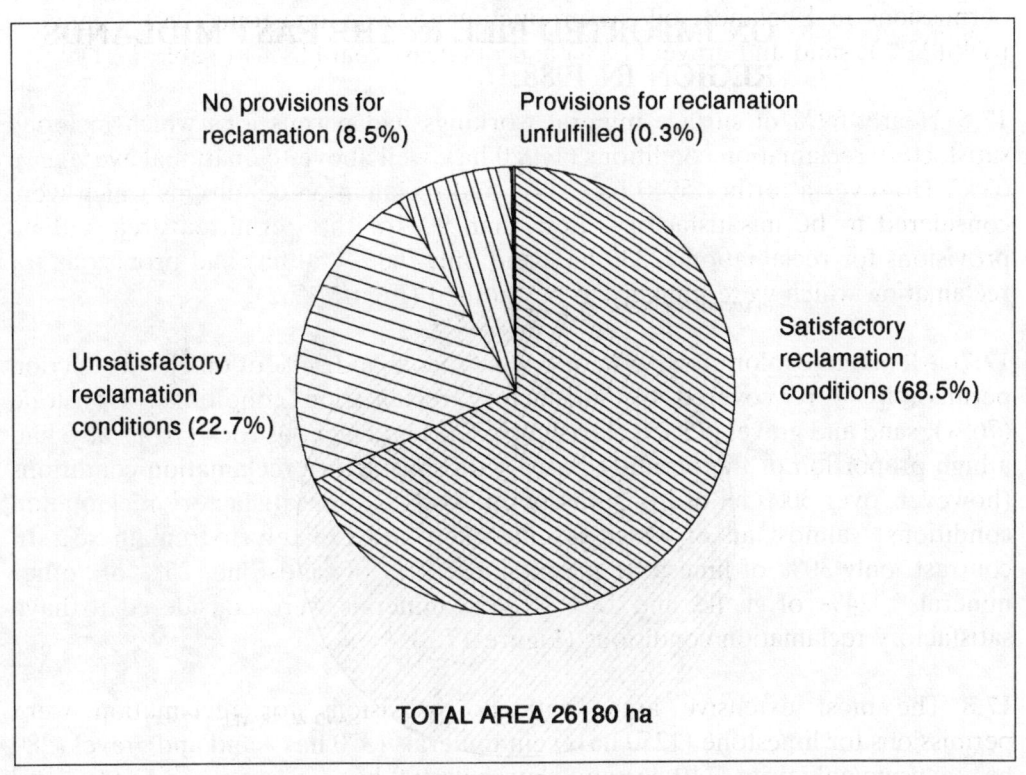

FIGURE 17.3 AREA AND NATURE OF PROVISIONS FOR RECLAMATION OF SURFACE MINERAL WORKINGS IN THE EAST MIDLANDS REGION IN 1988 – MAIN MINERAL TYPES.

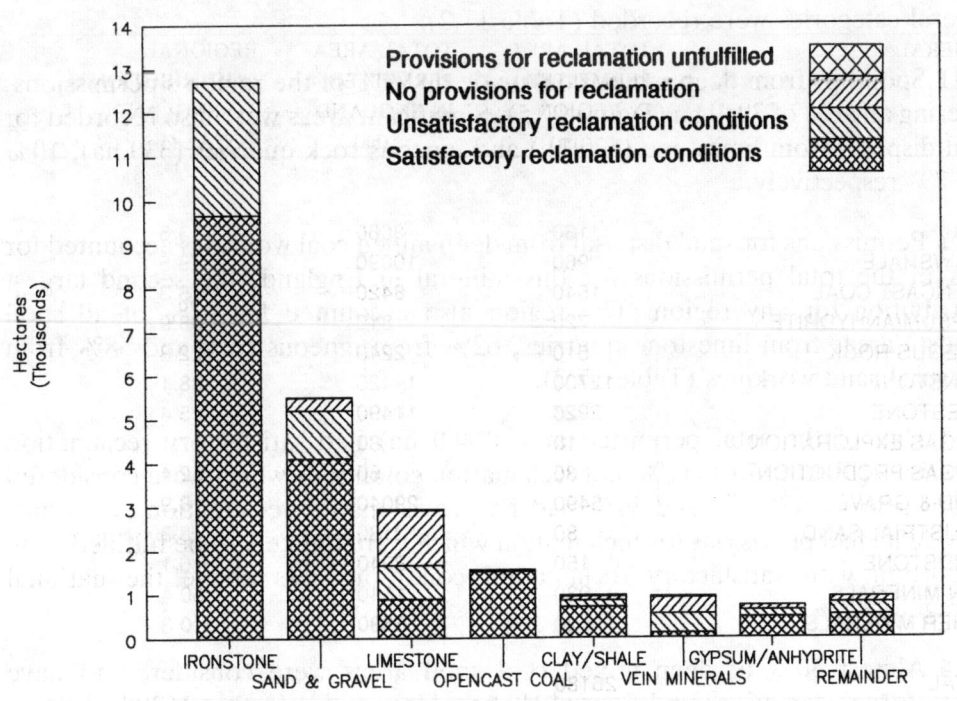

17.9 There were 1990 ha of workings which required fill to achieve reclamation, 7% of the permitted area in the region. Sand and gravel made up 77% of this area (1470 ha), clay/shale 6% (110 ha) and limestone 5% (90 ha) (Figure 17.4).

FIGURE 17.4 SURFACE MINERAL WORKINGS DEPENDENT ON IMPORTED FILL IN THE EAST MIDLANDS REGION IN 1988.

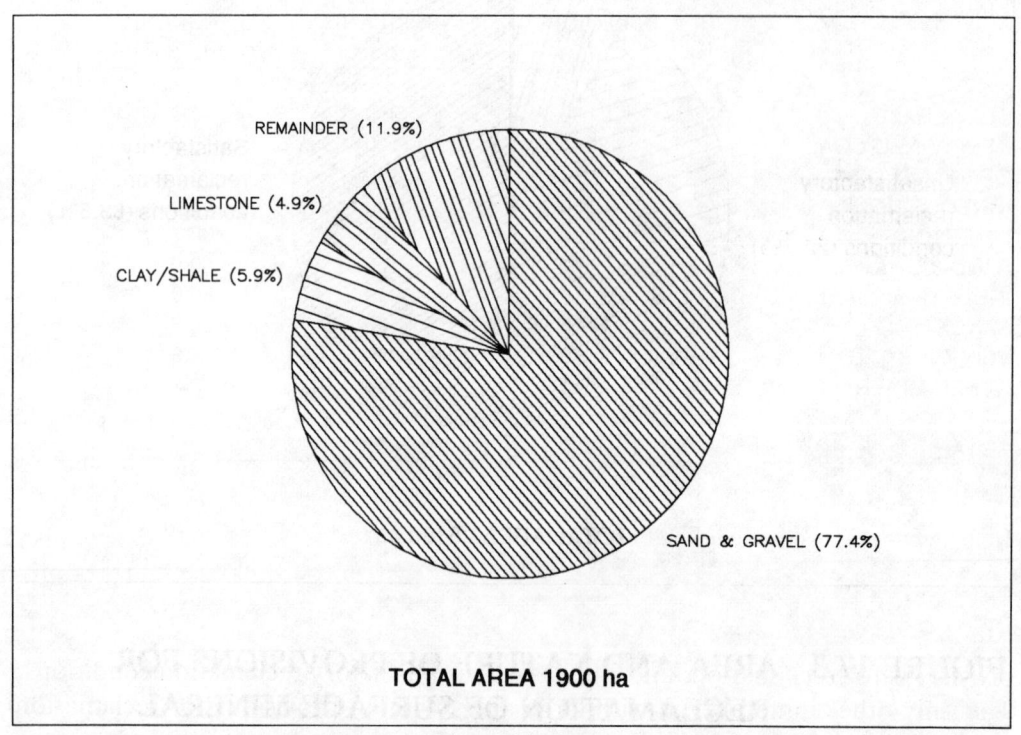

Surface Disposal of Mineral Working Deposits

17.10 The East Midlands region had the second largest permitted area for the surface disposal of mineral working deposits (ie spoil tips) in England, with permissions covering a total area of 4960 ha, 28% of the England total. Ten mineral categories were recorded (Table 17.2).

17.11 Spoil tips from deep mined coal made up 79% of the region's permissions, covering an area of 3930 ha (Figure 17.5). Significant areas were also recorded for spoil disposal from limestone (520 ha) and igneous rock quarries (330 ha), 10% and 7% respectively.

17.12 Permissions for spoil disposal from deep mined coal workings accounted for 35% of the total permissions for this mineral in England, the second largest contribution for any region. The region also accounted for 67% of all spoil disposal areas from limestone quarries, 62% from igneous rock and 58% from industrial sand workings (Table 17.2).

17.13 Half of the total permitted area (2460 ha) had satisfactory reclamation conditions (Figure 17.6), 11% had reclamation conditions which were considered unsatisfactory (560 ha), and 39% had no provisions for reclamation (1920 ha). Only 20 ha had provisions for reclamation which were unlikely to be fulfilled. The proportion with satisfactory reclamation conditions was above the national average of 41%.

17.14 Almost 50% of deep mined coal spoil areas were considered to have satisfactory reclamation conditions (1940 ha). However, a further 1630 ha had no

TABLE 17.2 TOTAL AREA PERMITTED FOR THE SURFACE DISPOSAL OF MINERAL WORKING DEPOSITS IN THE EAST MIDLANDS REGION, INDICATING THE PERCENTAGE REGIONAL CONTRIBUTION TO THE ENGLAND TOTAL.

MINERAL	TOTAL AREA PERMITTED IN REGION (ha)	TOTAL AREA PERMITTED IN ENGLAND (ha)	REGIONAL CONTRIBUTION TO ENGLAND TOTAL (%)
CLAY/SHALE	30	350	8.6
DEEP MINED COAL	3930	11120	35.3
OPENCAST COAL	<10	130	3.1
IGNEOUS ROCK	330	530	62.3
LIMESTONE	520	780	66.7
SAND & GRAVEL	70	140	50.0
INDUSTRIAL SAND	10	20	50.0
SANDSTONE	20	60	33.3
VEIN MINERALS	50	320	15.6
OTHER MINERALS	<10	40	2.5
TOTAL	4960		

Note: Areas are rounded to the nearest 10 ha.

provisions for reclamation, and 340 ha had unsatisfactory reclamation conditions. The only other mineral type with large areas without provisions for reclamation was limestone (220 ha), 43% of the total permitted area for this mineral.

Underground mining

17.15 The extent of underground mining recorded in the region amounted to 195930 ha. This was the second largest area in England (after the Northern region). The vast majority of the area (93%) was coal mines. Most of the coal mines (83%), were worked under the GDO (151220 ha), while 17% (31230 ha) were worked under specific planning permissions. Permissions for mining gypsum/anhydrite (6620 ha), vein minerals (5080 ha) and ironstone (1360 ha) were also extensive.

17.16 All permissions for the underground mining of ironstone in England were within the East Midlands (Table 17.3). The region also had the largest permitted areas recorded in England for coal worked under the GDO (48% of the England total), coal worked under specific planning permissions (67%), and gypsum/anhydrite (45%).

Land covered by permissions which include aftercare conditions

17.17 There were 158 permissions which included aftercare conditions (144 for surface mineral workings and 14 for spoil tip areas) covering a total area of 4210 ha (3320 ha for surface mineral workings, 890 ha for spoil tips). This represented 13% of the current permitted area of surface workings and 18% of spoil tips.

17.18 Three mineral types accounted for 85% of the total area covered by aftercare conditions. Sand and gravel had the largest area (1500 ha) 36% of the region's total, followed by opencast coal (32% or 1350 ha) and spoil tips from

FIGURE 17.5 AREA PERMITTED FOR THE SURFACE DISPOSAL OF MINERAL WORKING DEPOSITS IN THE EAST MIDLANDS REGION IN 1988.

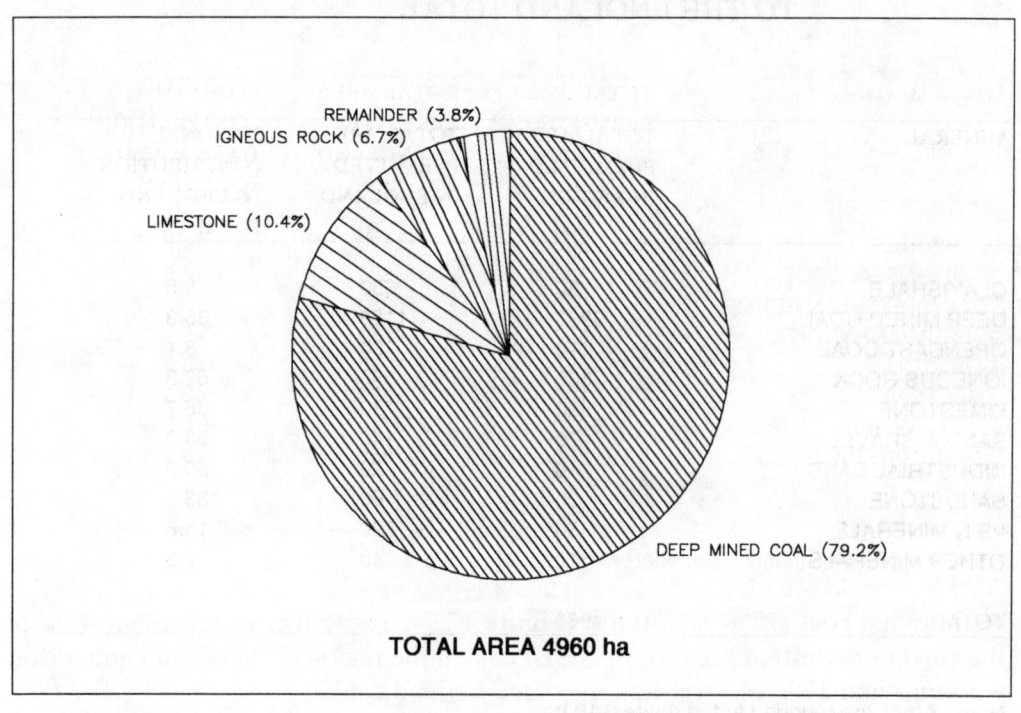

FIGURE 17.6 NATURE OF PROVISIONS FOR RECLAMATION FOR THE SURFACE DISPOSAL OF MINERAL WORKING DEPOSITS IN THE EAST MIDLANDS REGION IN 1988.

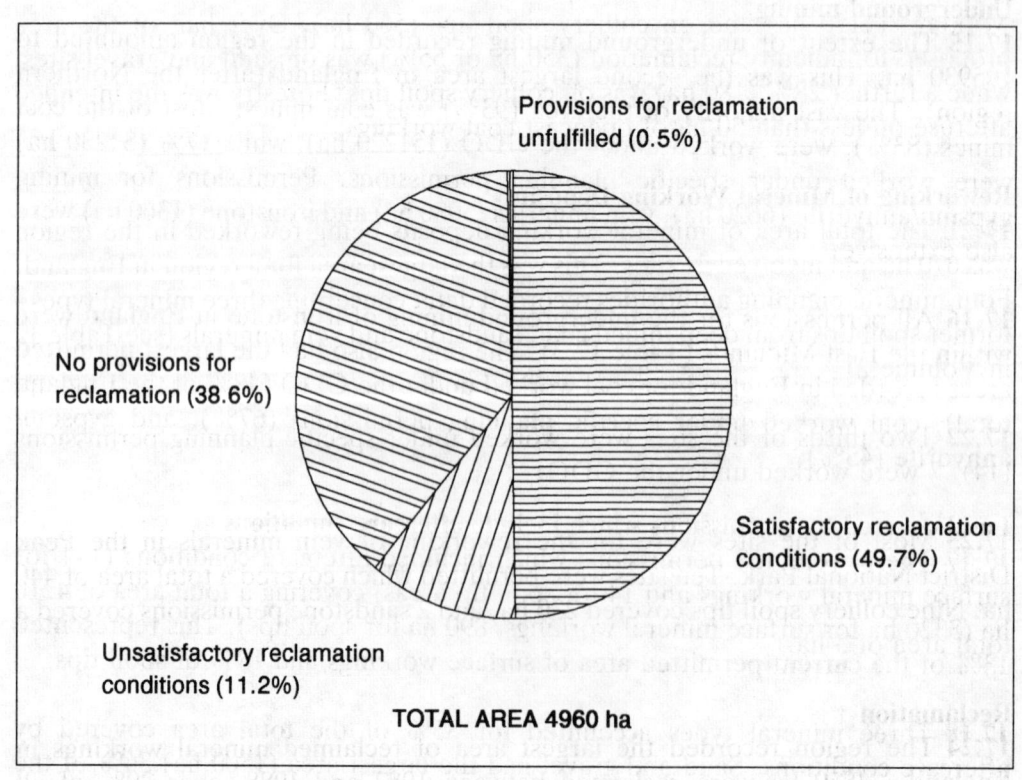

TABLE 17.3 TOTAL AREA OF UNDERGROUND PERMISSIONS IN THE EAST MIDLANDS REGION, INDICATING THE PERCENTAGE REGIONAL CONTRIBUTION TO THE ENGLAND TOTAL.

MINERAL	TOTAL AREA PERMITTED IN REGION (ha)	TOTAL AREA PERMITTED IN ENGLAND (ha)	REGIONAL CONTRIBUTION TO ENGLAND TOTAL (%)
CLAY/SHALE	290	1470	19.7
COAL (UNDER GDO)	151220	315820	47.9
COAL (SPEC PLAN PERM)	31230	46380	67.3
GYPSUM/ANHYDRITE	6620	14670	45.1
IRONSTONE	1360	1360	100.0
LIMESTONE	130	300	43.3
VEIN MINERAL	5080	379710	1.3
OTHER MINERALS	<10	17850	<0.1
TOTAL	195930		

Note: Areas are rounded to the nearest 10 ha.

deep mined coal (17% or 730 ha) (Figure 17.7). These figures represent 87% of the current permitted area for opencast coal in the region, 27% of sand and gravel workings and 19% of spoil tips from deep mined coal.

17.19 Reclamation to agriculture was proposed for 84% of the area with aftercare conditions (3530 ha). Amenity afteruses were proposed for 15% (630 ha) and forestry 1% (50 ha) (Figure 17.7).

17.20 All mineral types with aftercare conditions had an agricultural afteruse intended for at least some of the permitted area. The largest areas proposed for agricultural reclamation were on opencast coal sites (1250 ha), sand and gravel workings (1150 ha) and on colliery spoil tips (560 ha). Over half of the area proposed for amenity reclamation (350 ha or 55%) was on sand and gravel sites, while a further 28% (180 ha) was on colliery spoil tips. Forestry was the intended afteruse on less than 50 ha of opencast coal workings.

Reworking of Mineral Working Deposits

17.21 The total area of mineral working deposits being reworked in the region was 670 ha and involved 21 sites. This was the largest area for a region in England. Four mineral planning authorities recorded data, concerning three mineral types - former spoil tips from deep mined coal, sandstone and vein minerals (see Table J2 in Volume 2).

17.22 Two thirds of the sites were worked under specific planning permissions (14), 7 were worked under the GDO.

17.23 Most of the sites were for the reworking of vein minerals in the Peak District National Park. Ten sites were permitted which covered a total area of 440 ha. Nine colliery spoil tips covered 230 ha, and 2 sandstone permissions covered a total area of 3 ha.

Reclamation

17.24 The region recorded the largest area of reclaimed mineral workings in England - 4850 ha were reclaimed between 1982 and 1988, over 20% of all reclamation in England.

FIGURE 17.7 AREA OF MINERAL WORKINGS COVERED BY AFTERCARE CONDITIONS IN THE EAST MIDLANDS REGION IN 1988 – MAIN MINERAL TYPES, AND PROPOSED ENDUSES.

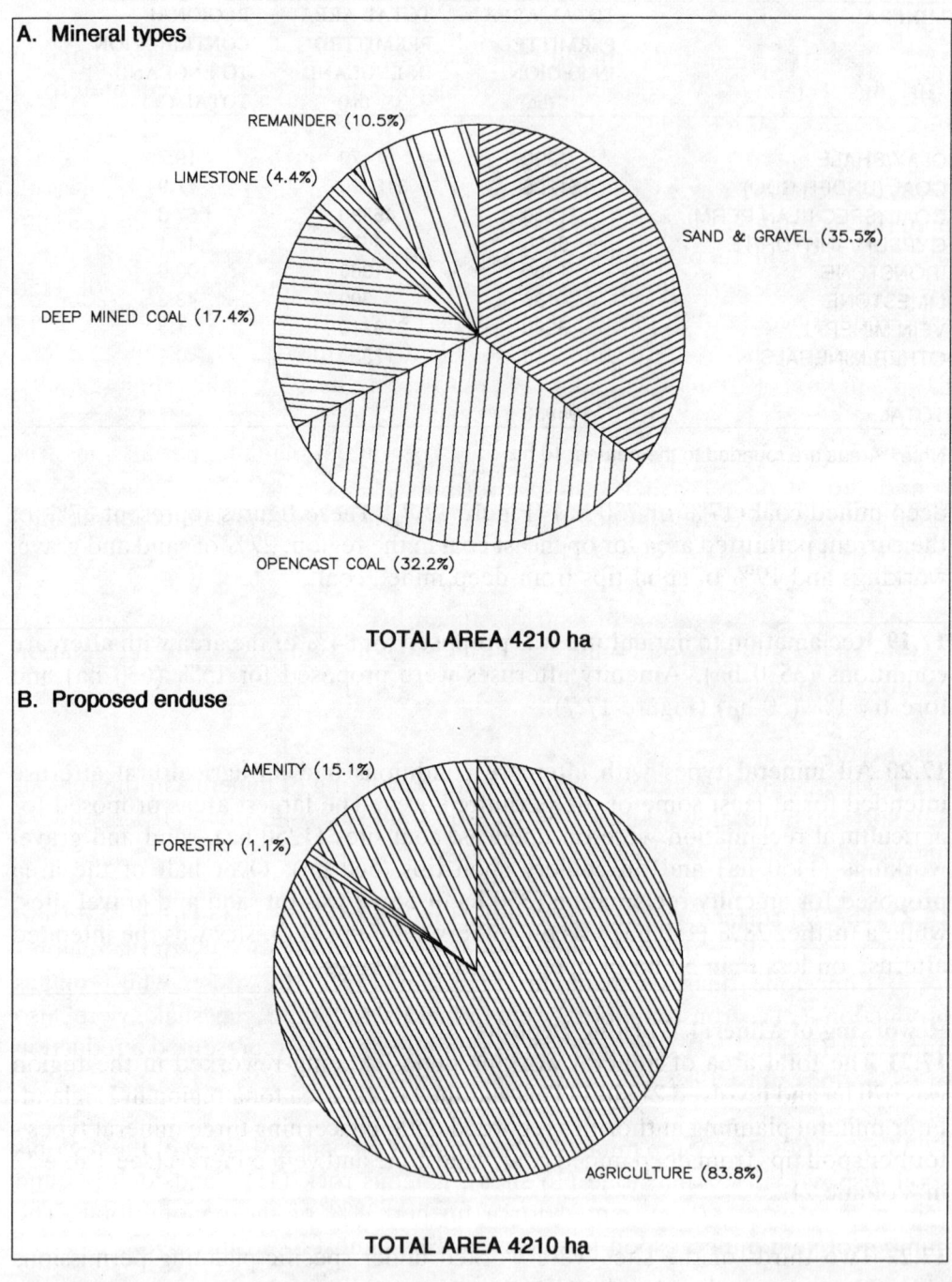

17.25 Reclamation as a result of mineral planning permissions accounted for 87% of the total (4210 ha). A further 640 ha were reclaimed by "other means".

17.26 The reclamation of sand and gravel workings accounted for 43% (1790 ha) of land reclaimed by mineral planning permissions. Spoil tips from deep mined coal accounted for a further 18% (750 ha), opencast coal 13% (560 ha) and ironstone 12% (480 ha) (Figure 17.8).

17.27 Spoil tips from deep mined coal accounted for 46% (290 ha) of mineral workings reclaimed by "other means". Sand and gravel workings accounted for a further 28% (180 ha) (Figure 17.8).

17.28 Agriculture was the most common afteruse for land reclaimed by both mineral planning permissions and "other means". Almost 67% (2810 ha) of land reclaimed by mineral planning permissions, and 38% (240 ha) reclaimed by "other means" was to agriculture. Amenity afteruses accounted for 30% of land reclaimed by mineral planning permissions (1260 ha), while 2% (90 ha) were reclaimed to "other uses". Almost 33% of land reclaimed by "other means" was to amenity uses (210 ha), while 27% were to "other uses" (170 ha) (Figure 17.9).

17.29 Ninety six percent of land reclaimed by mineral planning permissions was considered satisfactory, slightly above the national average (95%). Vein minerals and spoil tips from deep mined coal had the largest areas reclaimed unsatisfactorily (60 ha and 50 ha respectively).

17.30 Ninety seven percent of land reclaimed by "other means" was considered satisfactory. Only clay/shale and limestone sites had unsatisfactory reclamation covering a total area of 30 ha.

Number of Sites
17.31 There were 912 sites associated with the working of minerals in the region, 721 surface mineral workings (79%), 110 spoil disposal areas (12%) and 81 underground workings (9%).

17.32 The 721 surface workings accounted for 21% of the England total. Sand and gravel workings were the most numerous (175) accounting for 24% of the region's total. Limestone quarries made up a further 15% (106 sites), whilst oil/gas production (77), ironstone (76), vein minerals (68) and clay/shale were also numerous (Table 17.4). The number of vein mineral sites represented a reduction of 69% (153) since 1982.

17.33 Spoil tips from deep mined coal accounted for 55% (61 sites) of the region's spoil disposal sites. Limestone (16 sites), igneous rock (11), sandstone (8) and vein mineral sites (7) accounted for a further 38% of the regions total. The number of vein mineral spoil tips decreased by almost 87% since 1982.

17.34 Coal and vein mineral mines were the most numerous underground workings in the region with 44 and 23 mines respectively (Table 17.4).

Land included in the Derelict Land Survey
17.35 There were 680 ha of derelict mineral workings included in both this and the Derelict Land Survey, 16% of the England total.

17.36 Areas affected by spoil tips from deep mined coal (490 ha) accounted for 72% of the derelict land in the region. Sand and gravel (60 ha), limestone (60 ha), and clay/shale (40 ha) sites were also recorded as derelict (Figure 17.10).

FIGURE 17.8 AREA OF LAND RECLAIMED BY MINERAL PLANNING PERMISSIONS (A) AND BY OTHER MEANS (B) IN THE EAST MIDLANDS REGION BETWEEN 1982 AND 1988.

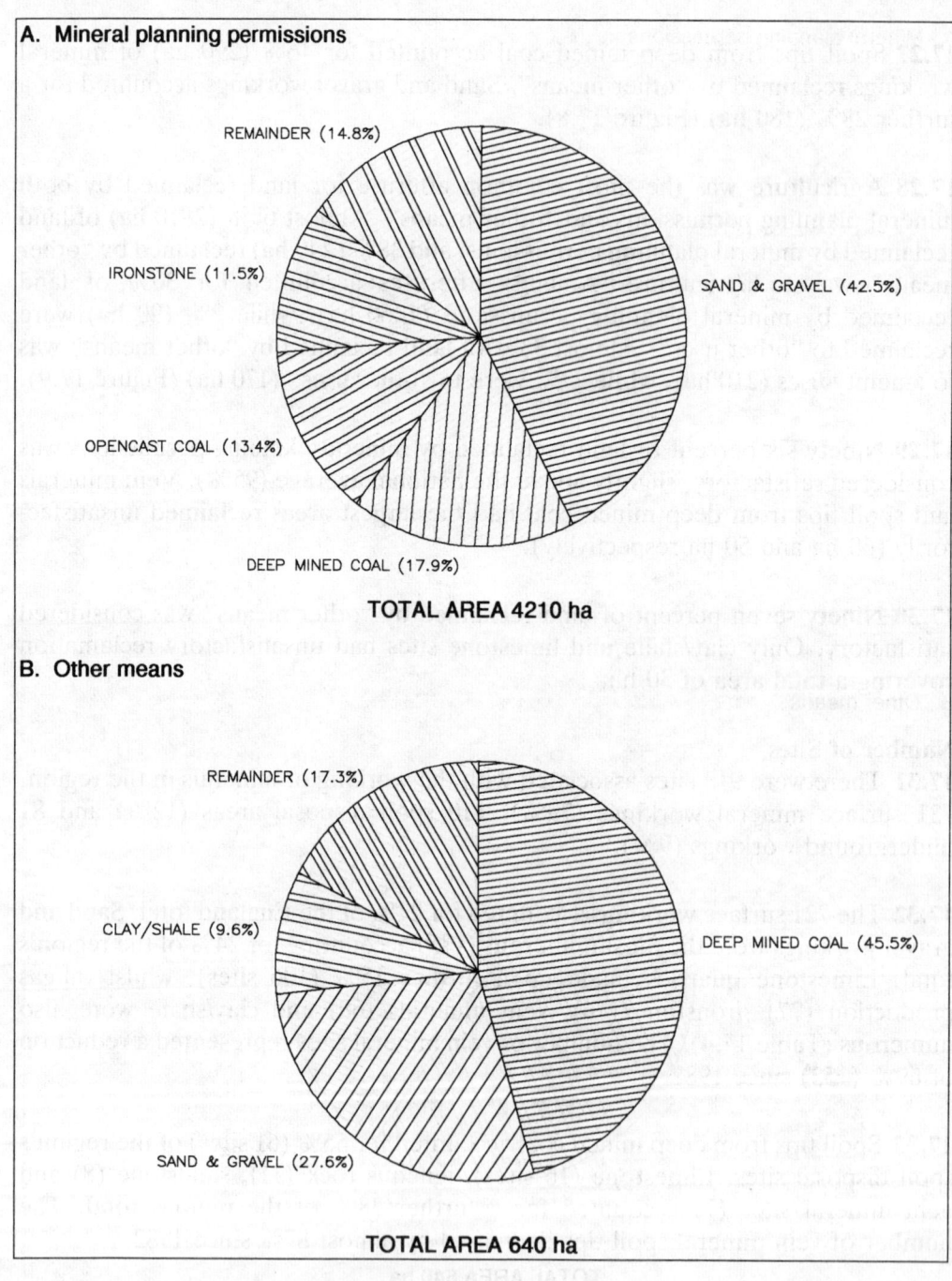

FIGURE 17.9 CHOSEN AFTERUSE OF LAND RECLAIMED BY MINERAL PLANNING PERMISSIONS (A) AND BY OTHER MEANS (B) IN THE EAST MIDLANDS REGION BETWEEN 1982 AND 1988.

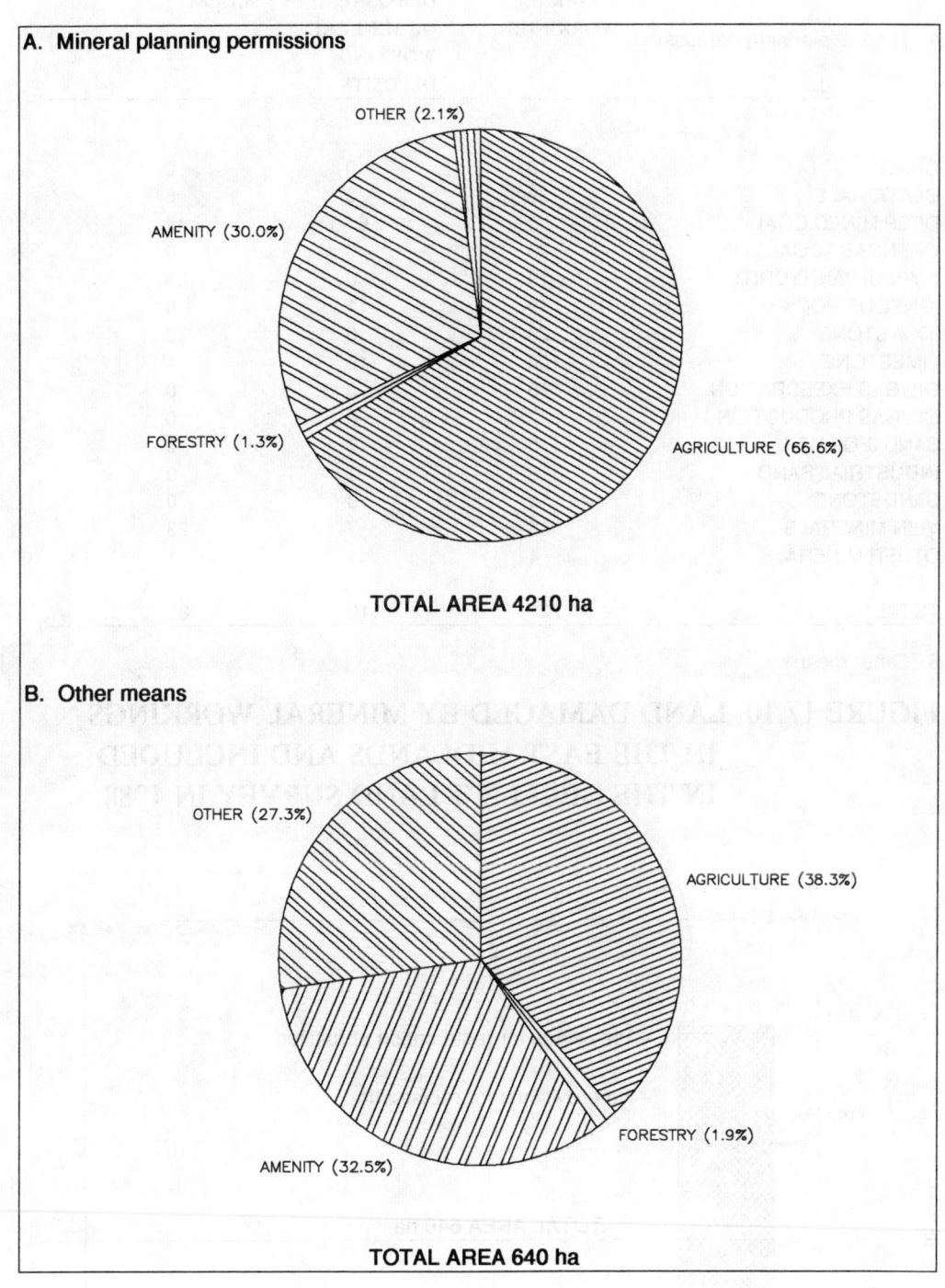

TABLE 17.4 NUMBER OF SITES FOR MINERAL WORKINGS IN THE EAST MIDLANDS REGION IN 1988.

MINERAL	SURFACE MINERAL WORKINGS	SURFACE DISPOSAL OF MINERAL WORKING DEPOSITS	UNDERGROUND MINING
CHALK	10	0	0
CLAY/SHALE	59	3	6
DEEP MINED COAL	0	61	44
OPENCAST COAL	29	0	0
GYPSUM/ANHYDRITE	19	0	3
IGNEOUS ROCK	15	11	0
IRONSTONE	76	0	3
LIMESTONE	106	16	1
OIL/GAS EXPLORATION	15	0	0
OIL/GAS PRODUCTION	77	0	0
SAND & GRAVEL	175	2	0
INDUSTRIAL SAND	15	1	0
SANDSTONE	53	8	0
VEIN MINERALS	68	7	23
OTHER MINERALS	4	1	1
TOTAL	721	110	81

FIGURE 17.10 LAND DAMAGED BY MINERAL WORKINGS IN THE EAST MIDLANDS AND INCLUDED IN THE DERELICT LAND SURVEY IN 1988.

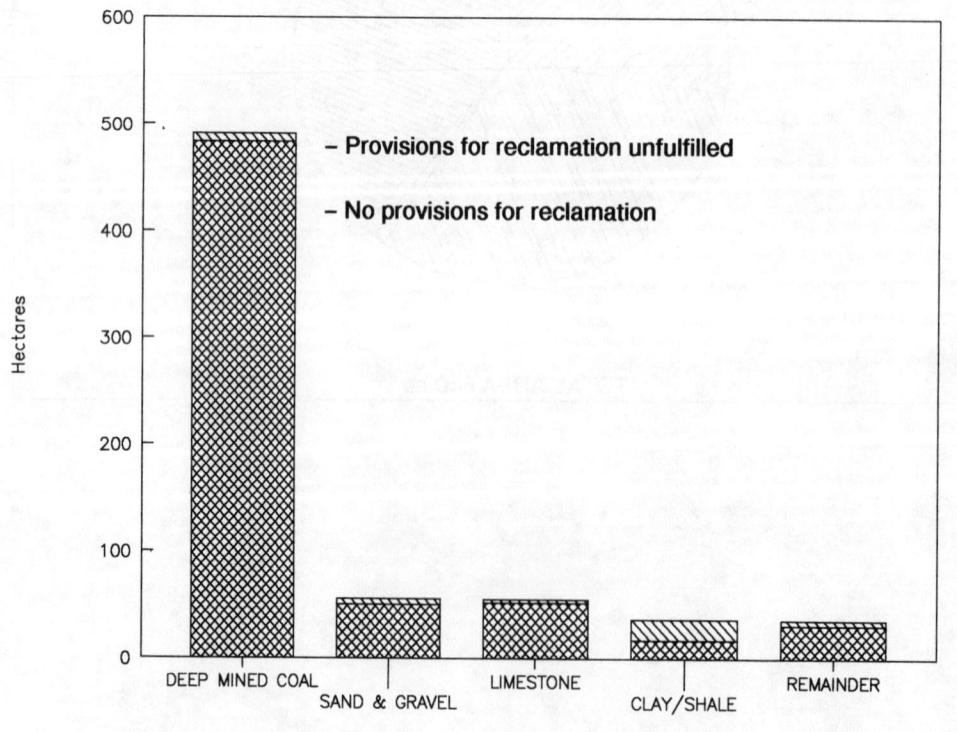

Reviews under Section 264A of the Town and Country Planning Act

17.37 Two mineral planning authorities within the region had initiated their review procedures by 1st April 1988, Derbyshire and Leicestershire.

17.38 Reviews had been carried out of 113 sites and 5 unimplemented permissions. Only 12 sites had been identified for further action, 10% of those reviewed. Two voluntary agreements had been made, but no other action had been identified.

17.39 Most of the sites identified for further action were clay/shale workings, of which 8 out of the 13 reviewed were identified for further action.

18. EAST ANGLIA

Introduction

18.1 East Anglia contains three mineral planning authorities; Cambridgeshire, Norfolk and Suffolk. Mineral workings were recorded by all authorities.

18.2 There are three AONB's in the region; the Norfolk Coast AONB in Norfolk, the Suffolk Coast and Heaths AONB and the Dedham Vale AONB in Suffolk. Surface mineral workings were recorded in all three AONB's.

Surface Mineral Workings

18.3 East Anglia had the second smallest area of permissions for surface mineral workings, and the smallest number of mineral categories worked in England. The total permitted area was 5460 ha, less than 6% of the England total. Seven mineral categories (including "other minerals") were recorded.

18.4 Sand and gravel permissions were the most extensive accounting for 57% of the total area (3110 ha). Clay/shale permissions the second most widespread mineral type, accounted for 12% of the recorded area (1210 ha), while industrial sand (470 ha) and chalk (350 ha) made up a further 9% and 6% respectively (Figure 18.1).

18.5 Despite the predominance of sand and gravel permissions in the region, these represented less than 11% of the total permitted area for the mineral in England (Table 18.1). Permissions for the extraction of industrial sand accounted for 22% of the total permitted area for the mineral in England, whilst clay/shale and chalk permissions accounted for 12% and 10% respectively.

FIGURE 18.1 AREA OF SURFACE MINERAL WORKINGS PERMITTED IN THE EAST ANGLIA REGION IN 1988 – MAIN MINERAL TYPES.

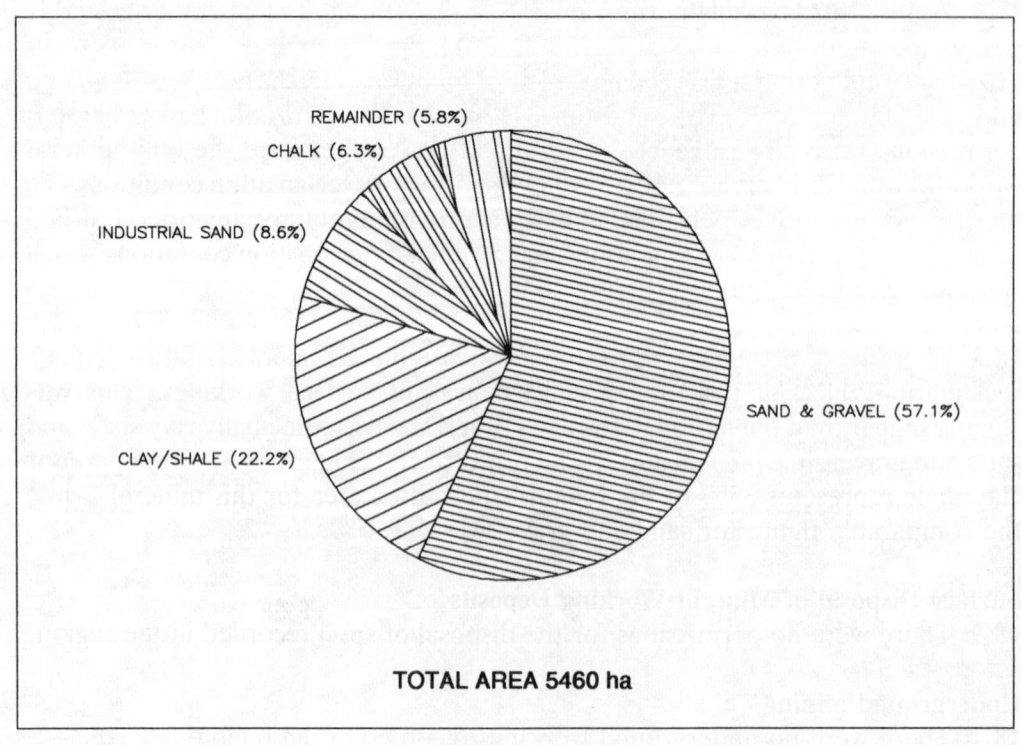

TABLE 18.1 TOTAL AREA PERMITTED FOR SURFACE MINERAL WORKINGS IN THE EAST ANGLIA REGION, INDICATING THE PERCENTAGE REGIONAL CONTRIBUTION TO THE ENGLAND TOTAL.

MINERAL	TOTAL AREA PERMITTED IN REGION (ha)	TOTAL AREA PERMITTED IN ENGLAND (ha)	REGIONAL CONTRIBUTION TO ENGLAND TOTAL (%)
CHALK	350	3660	9.6
CLAY/SHALE	1210	10090	12.0
LIMESTONE	160	11490	1.4
SAND & GRAVEL	3110	29040	10.7
INDUSTRIAL SAND	470	2130	22.1
SANDSTONE	130	2940	4.4
OTHER MINERALS	30	6530	0.5
TOTAL	5460		

Note: Areas are rounded to the nearest 10 hectares.

18.6 Over 94% of the permitted area (5160 ha) had satisfactory reclamation conditions (Figure 18.2). This was by far the highest regional percentage for permissions with satisfactory reclamation conditions, well above the national average of 63%. Only 150 ha (2%) had unsatisfactory reclamation conditions, 1% (80 ha) had no provisions for reclamation and 1% (80 ha) had been damaged and was unlikely to be reclaimed because provisions for reclamation remained unfulfilled.

18.7 Almost all clay/shale permissions (99% or 1210 ha) and all "other mineral" permissions (30 ha) had satisfactory reclamation conditions, while 98% of industrial sand permissions (460 ha), 95% of limestone (150 ha), 93% of sand and gravel (2900 ha), 93% of sandstone (120 ha) and 87% of chalk (300 ha) permissions were also satisfactory. Sand and gravel sites were the only mineral workings with sizable areas not covered by satisfactory reclamation conditions - 50 ha of sand and gravel permissions had no provisions for reclamation, 120 had unsatisfactory reclamation conditions and 50 ha had reclamation conditions which were unlikely to be fulfilled (Figure 18.3).

18.8 There were 620 ha of permissions which required imported fill to achieve reclamation, 11% of the total permitted area of surface workings. This was slightly higher than the national average (10%), and was principally clay/shale and sand and gravel sites (430 ha and 150 ha respectively) (Figure 18.4). The figure for clay/shale represents 35% of the regions's permitted area for this mineral, while the comparable figure for sand and gravel was only 5%.

Surface Disposal of Mineral Working Deposits
18.9 There were no permissions for the disposal of spoil recorded in the region.

Underground mining
18.10 There were no underground workings recorded in the region.

FIGURE 18.2 NATURE OF PROVISIONS FOR RECLAMATION FOR SURFACE MINERAL WORKINGS IN THE EAST ANGLIA REGION IN 1988.

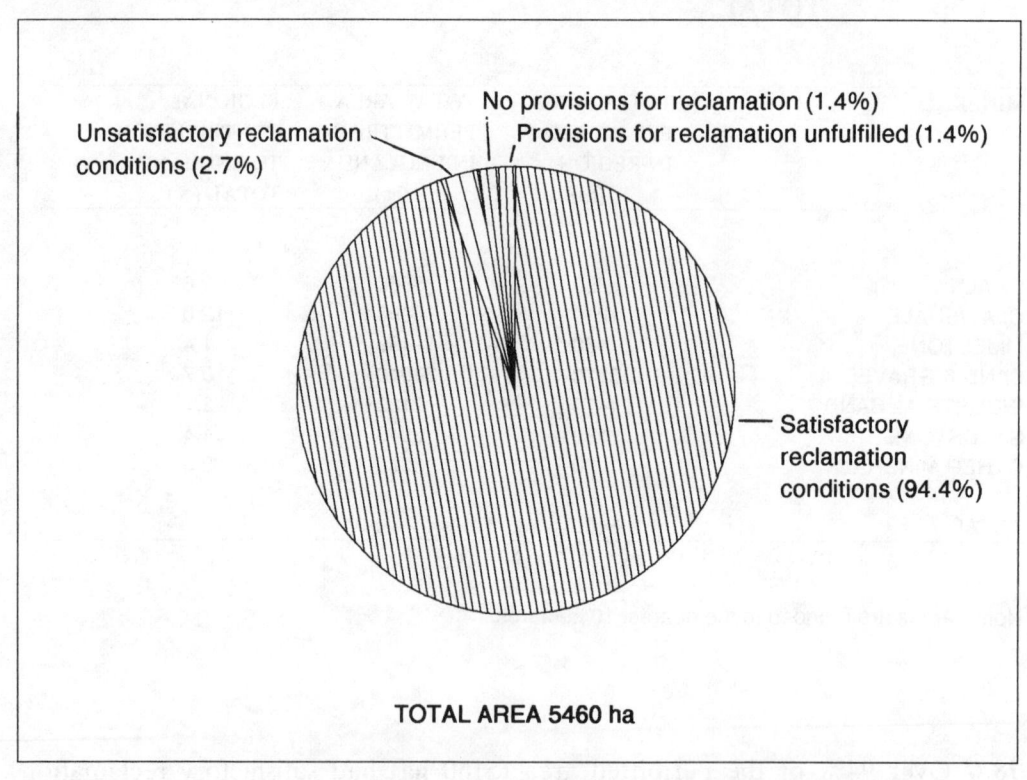

FIGURE 18.3 AREA AND NATURE OF PROVISIONS FOR RECLAMATION OF SURFACE MINERAL WORKINGS IN THE EAST ANGLIA REGION IN 1988 – MAIN MINERAL TYPES.

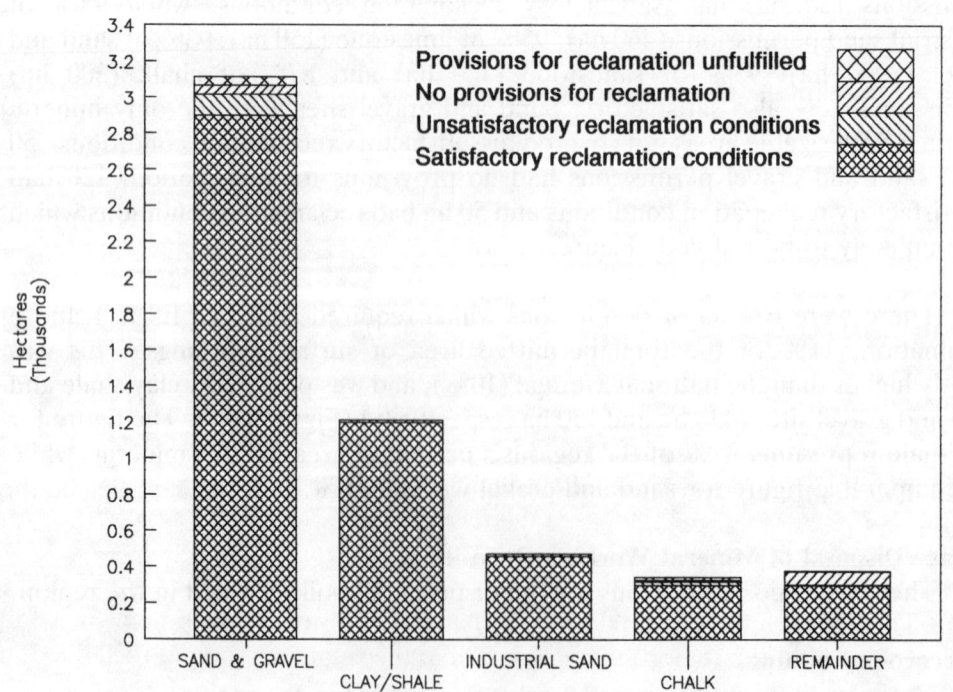

FIGURE 18.4 SURFACE MINERAL WORKINGS DEPENDENT ON IMPORTED FILL IN THE EAST ANGLIA REGION IN 1988 – MAIN MINERAL TYPES

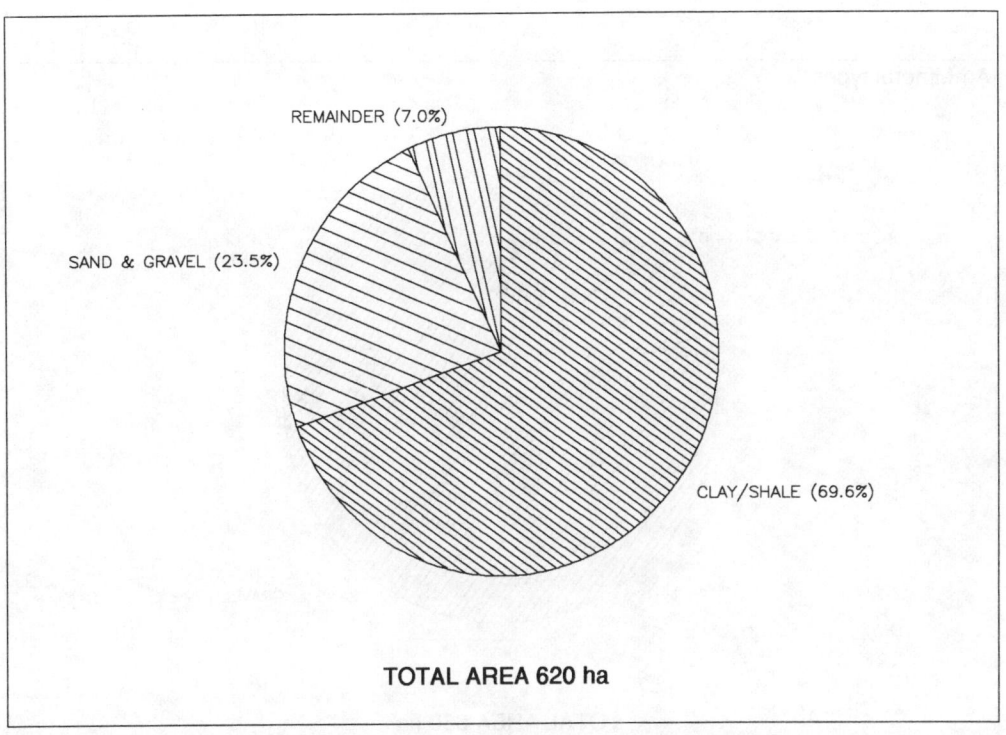

Land covered by permissions which include aftercare conditions
18.11 There were 74 permissions for surface mineral workings which included aftercare conditions, covering an area of 860 ha. This area is equivalent to almost 16% of the area of current permissions in the region. This was slightly lower than the England average of 18%.

18.12 Five mineral types had permissions which included aftercare conditions although sand and gravel permissions accounted for 80% of the total area (680 ha). Industrial sand accounted for a further 14% (120 ha) (Figure 18.5).

18.13 Agriculture was proposed as the final afteruse for 75% of the area with aftercare conditions (650 ha). Amenity uses accounted for a further 22% (190 ha) and forestry 3% (20 ha) (Figure 18.5). This contrasts with the balance of afteruses of land reclaimed since 1982 (see paragraph 18.19).

18.14 Since sand and gravel workings accounted for most of the land covered by aftercare conditions, it is not surprising that it accounted for 83% of the land to be reclaimed to agriculture and 74% of proposed amenity reclamation. However, less than 10 ha of sand and gravel permissions had forestry as the proposed afteruse with most of the land to be reclaimed to forestry (58%) being on industrial sand workings (14 ha).

Reworking of Mineral Working Deposits
18.15 There was no record of reworking of mineral working deposits in the region.

Reclamation
18.16 There were a total of 1100 ha of land affected by mineral working which had been reclaimed in the region since 1982, 5% of the total area of land reclaimed in England over that period.

FIGURE 18.5 AREA OF MINERAL WORKINGS COVERED BY AFTERCARE CONDITIONS IN THE EAST ANGLIA REGION IN 1988 – MAIN MINERAL TYPES, AND PROPOSED ENDUSES.

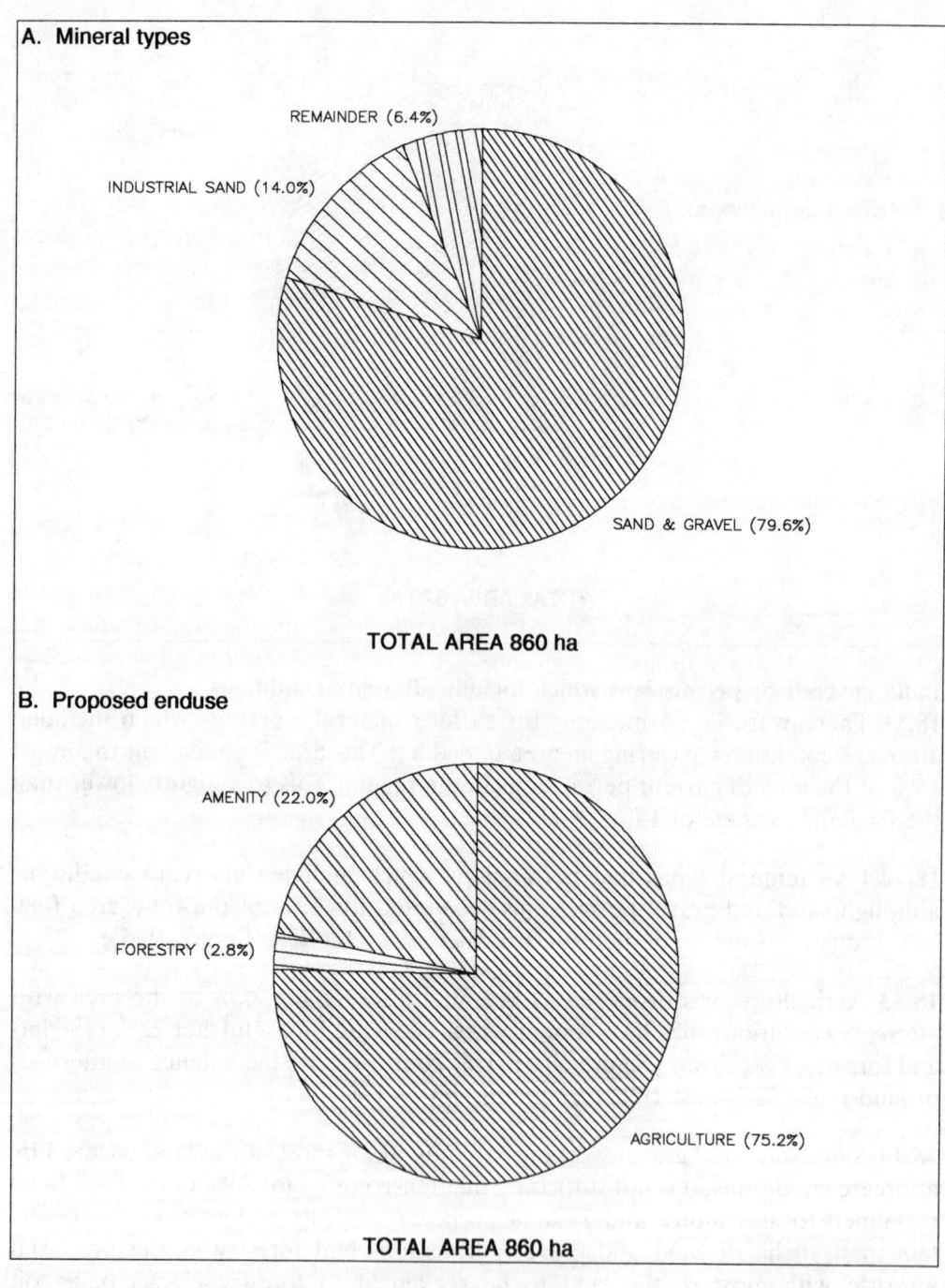

18.17 Over 94% of reclamation (1040 ha) was by mineral planning permissions. Only 60 ha (6%) were reclaimed by "other means". This was the highest ratio of land reclaimed by mineral planning permissions as opposed to "other means" in England.

18.18 Sand and gravel sites accounted for 78% (810 ha) of the land reclaimed by mineral planning permissions. Clay/shale (80 ha), industrial sand (50 ha) and chalk (50 ha) workings made up most of the remainder (Figure 18.6).

18.19 Only three mineral types were reclaimed by "other means", sand and gravel (40 ha), clay/shale (20 ha) and chalk (10 ha) (Figure 18.6).

18.20 Amenity afteruse was the most common for mineral workings reclaimed by mineral planning permissions, accounting for 60% of the total (620 ha) (Figure 18.7). This was one of only three regions where an agricultural afteruse was not the most common. Agriculture accounted for only 36% (380 ha) of the region's reclamation. Land reclaimed by "other means" was principally to agriculture (40 ha). The remaining 20 ha were reclaimed to "other uses" (Figure 18.7).

18.21 All of the land reclaimed by "other means" (60 ha) and 92% of land reclaimed by mineral planning permissions (950 ha) was considered satisfactory. Only small areas of sand and gravel (70 ha) and industrial sand (10 ha) workings were recorded as being reclaimed to an unsatisfactory standard.

Number of sites
18.22 East Anglia was the only region not to have sites with permissions for either the surface disposal of mineral working deposits or underground mines. All mineral sites (276) were surface workings. This was the smallest number of surface mineral workings recorded within a region, and represented only 6% of the England total.

18.23 There were 200 sites permitted for sand and gravel extraction, 72% of the region's total. Chalk (29 sites) and clay/shale (24 sites) were second and third most numerous (Table 18.2).

TABLE 18.2   NUMBER OF SITES FOR MINERAL WORKINGS IN THE EAST ANGLIA REGION IN 1988.

MINERAL	SURFACE MINERAL WORKINGS	SURFACE DISPOSAL OF MINERAL WORKING DEPOSITS	UNDERGROUND MINING
CHALK	29	0	0
CLAY/SHALE	24	0	0
LIMESTONE	7	0	0
SAND & GRAVEL	200	0	0
INDUSTRIAL SAND	4	0	0
SANDSTONE	7	0	0
OTHER MINERALS	5	0	0
TOTAL	276	0	0

FIGURE 18.6 AREA OF LAND RECLAIMED BY MINERAL PLANNING PERMISSIONS (A) AND BY OTHER MEANS (B) IN THE EAST ANGLIA REGION BETWEEN 1982 AND 1988 – MAIN MINERAL TYPES.

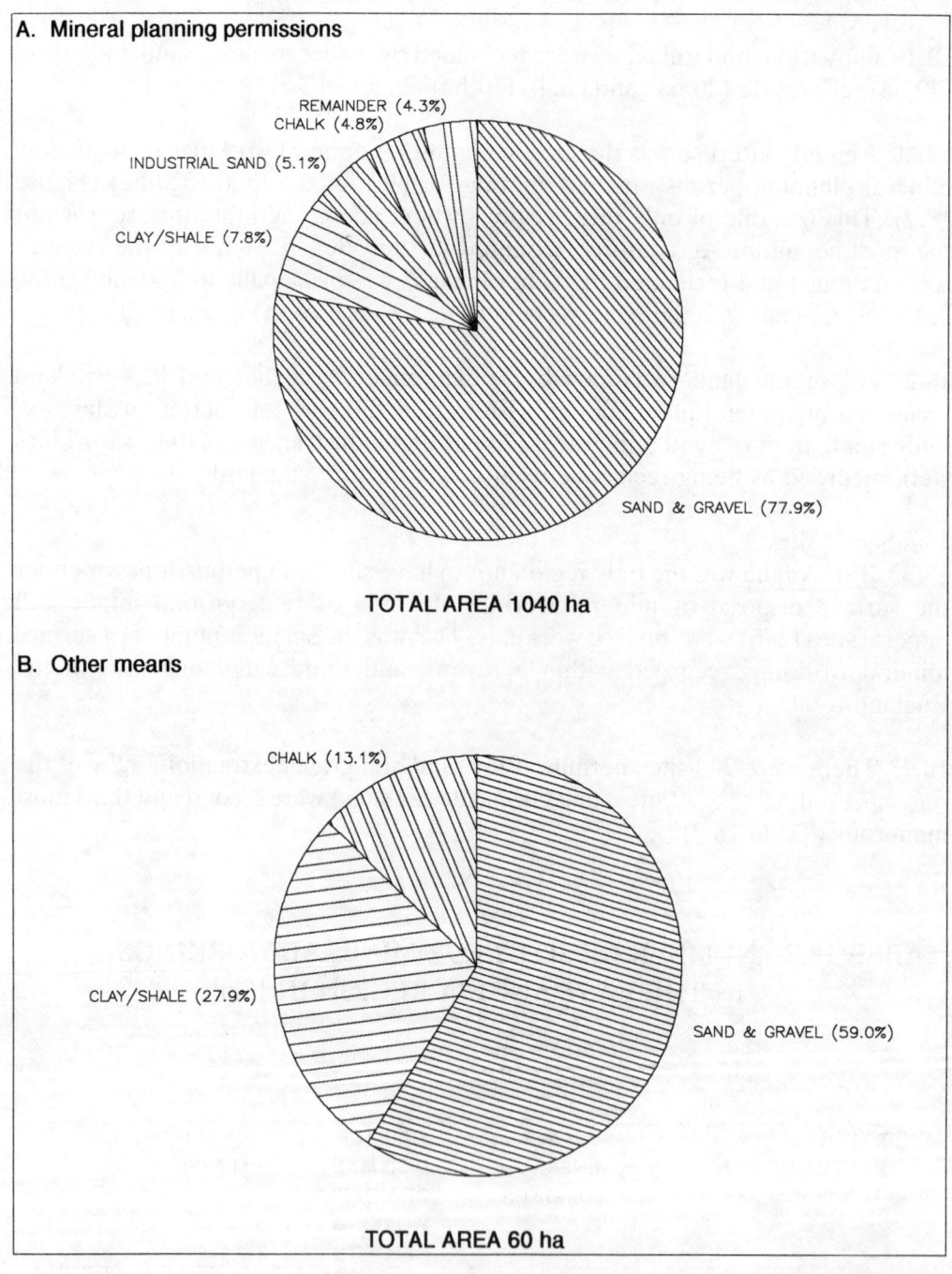

FIGURE 18.7 CHOSEN AFTERUSE OF LAND RECLAIMED BY MINERAL PLANNING PERMISSIONS (A) AND BY OTHER MEANS (B) IN THE EAST ANGLIA REGION BETWEEN 1982 AND 1988.

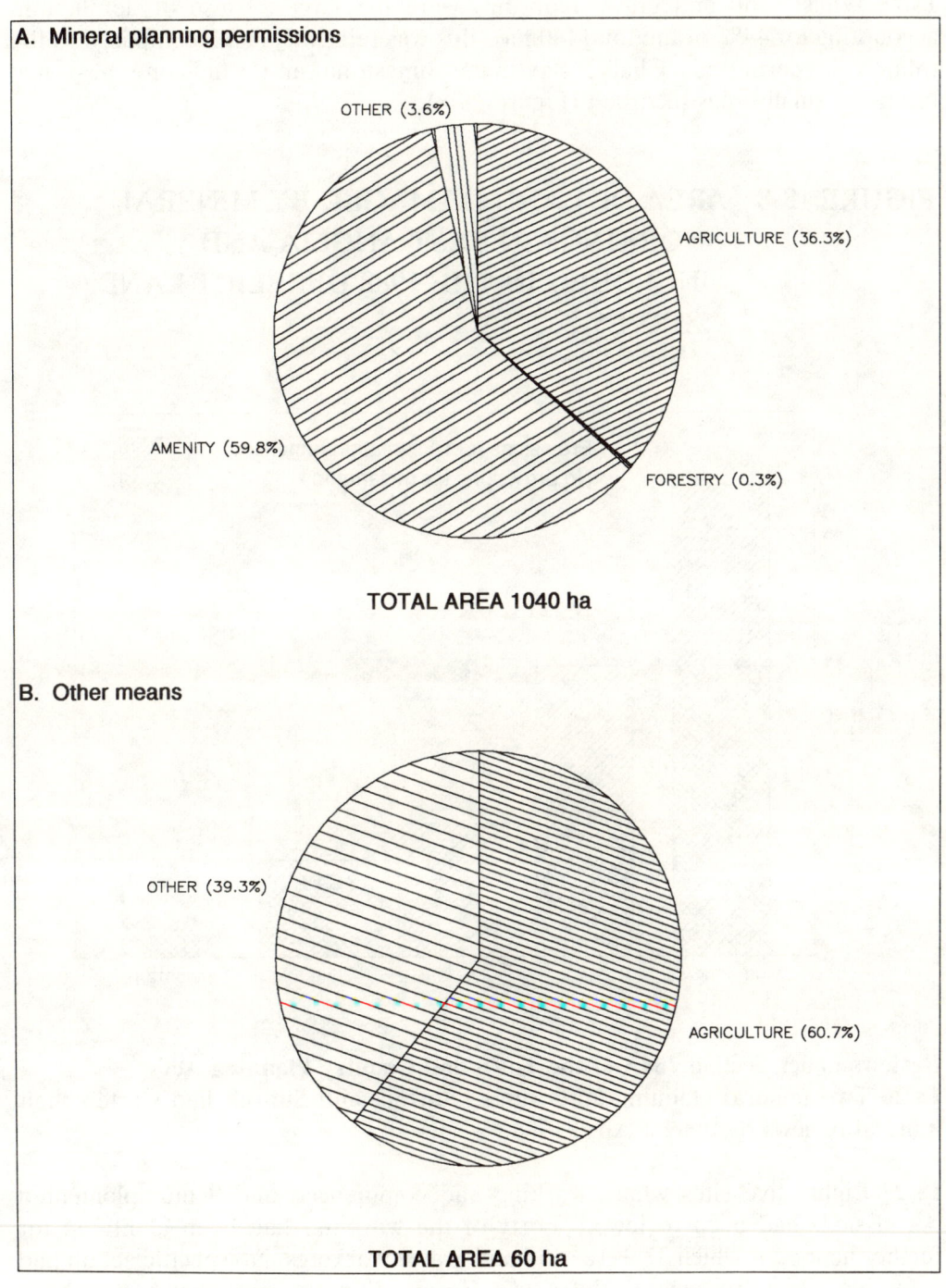

Land Included in the Derelict Land Survey

18.24 East Anglia had the smallest area of land which had become derelict as a result of mineral workings. Less than 90 ha of dereliction were recorded, less than 2% of the region's total permitted area. Approximately equal areas were derelict as a result of no provisions for reclamation (50 ha) and because provisions for reclamation were unfulfilled (40 ha).

18.25 Whilst sand and gravel workings were the largest cause of dereliction, accounting for 44% of the total (40 ha), this was relatively small compared to the total area permitted. Chalk, clay/shale, limestone and sandstone also had relatively small areas recorded (Figure 18.8).

FIGURE 18.8 AREA OF LAND DAMAGED BY MINERAL WORKINGS IN EAST ANGLIA AND INCLUDED IN THE 1988 DERELICT LAND SURVEY.

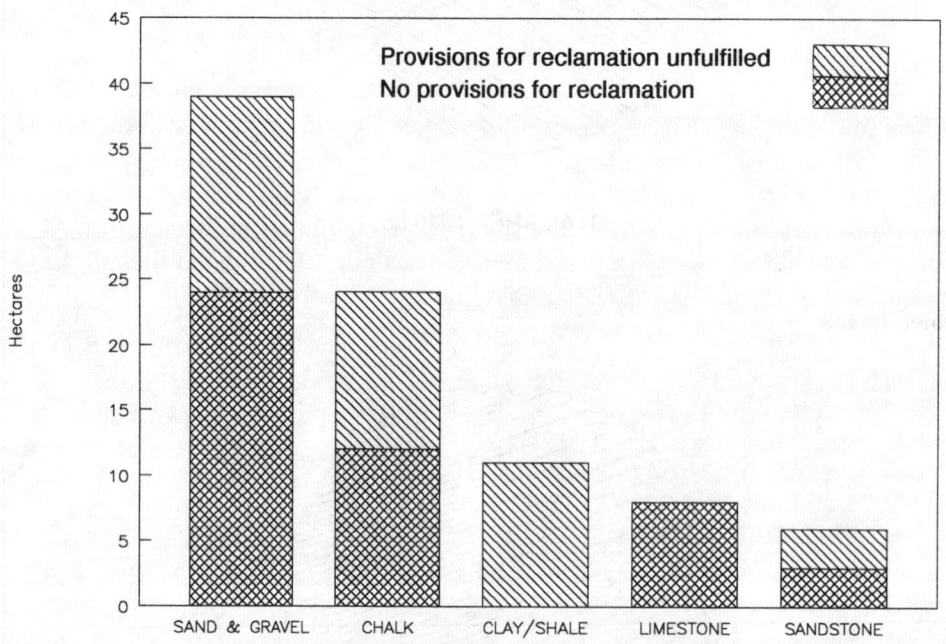

Reviews under Section 264A of the Town and Country Planning Act

18.26 Two mineral planning authorities, Norfolk and Suffolk had started their mineral reviews by the 1st April 1988.

18.27 Eighty five sites where working had commenced and 9 unimplemented permissions had been reviewed. Forty of the workings had been identified for further action, of which 37 were sand and gravel workings. No specific action had been identified for any of the sites. Three voluntary agreements had been achieved.

19. SOUTH EAST

Introduction

19.1 The South East region contains twelve shire counties; (Bedfordshire, Berkshire, Buckinghamshire, East Sussex, Essex, Hampshire, Hertfordshire, Isle of Wight, Kent, Oxfordshire, Surrey and West Sussex), and thirty three London Boroughs. Only seven of the London Boroughs recorded mineral workings within their areas, Bexley, Havering, Hillingdon, Hounslow, Merton, Redbridge and Sutton. Minerals were worked in all of the county areas.

19.2 There were 12 AONB's in the region. Minerals were worked in 9 of these as indicated in Table 19.1.

19.3 Oxfordshire were unable to provide data for this survey. The figures used for analysis are those provided by this authority for the 1982 Minerals Survey. Information is not available for this mpa about the amount of land covered by aftercare conditions, land restored since 1982, the extent of reworking of mineral working deposits and sites reviewed under Section 264A of the 1971 Act.

Surface Mineral workings

19.4 The South East had the second largest permitted area for surface mineral working in England. A total of 18510 ha were recorded, 19% of the overall permitted area for surface working in England.

19.5 Ten of the mineral categories used in the survey were recorded in the region. However, 90% of the total permitted area related to the extraction of three mineral types (Figure 19.1). Permissions for the extraction of sand and gravel were the most extensive, covering an area of 10880 ha and accounting for 59% of the area of surface workings in the region. Clay/shale (3500 ha) and chalk (2240 ha) permissions accounted for a further 19% and 12% respectively.

FIGURE 19.1 AREA OF SURFACE MINERAL WORKINGS PERMITTED IN THE SOUTH EAST REGION IN 1988 – MAIN MINERAL TYPES.

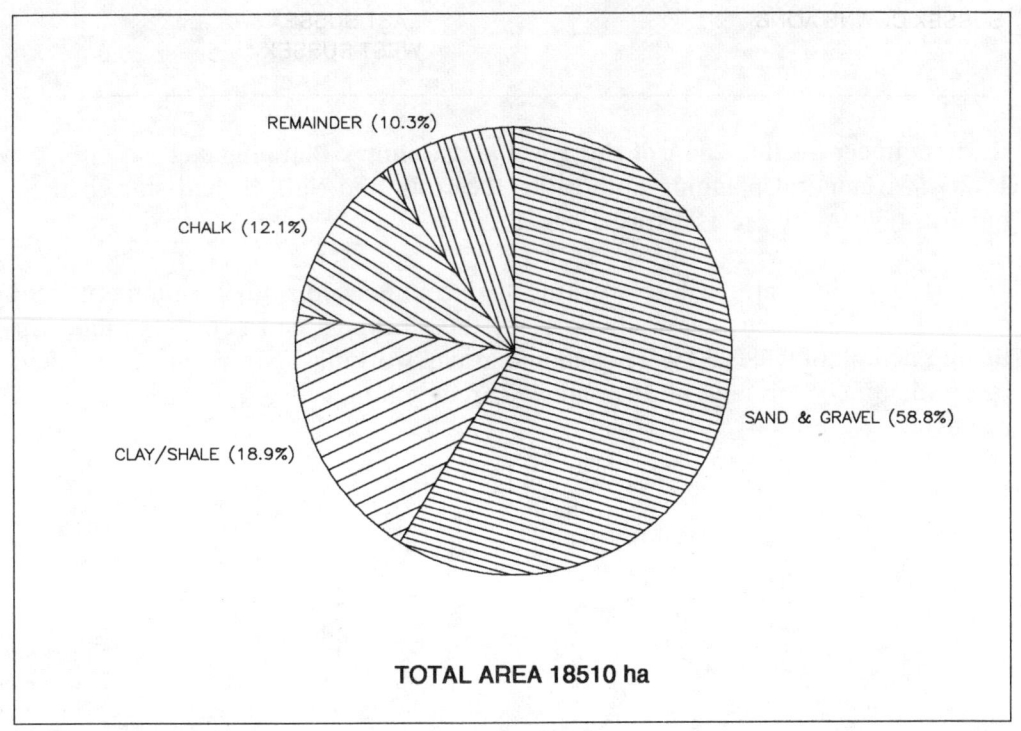

TABLE 19.1 AREAS OF OUTSTANDING NATURAL BEAUTY IN THE SOUTH EAST REGION, INDICATING RELEVANT MINERAL PLANNING AUTHORITIES AND THOSE WITH MINERAL WORKINGS (*).

AONB	MPA
CHICHESTER HARBOUR AONB	HAMPSHIRE WEST SUSSEX
CHILTERNS AONB	BEDFORDSHIRE * BUCKINGHAMSHIRE * HERTFORDSHIRE * OXFORDSHIRE *
COTSWOLDS AONB	OXFORDSHIRE
DEDHAM VALE AONB	ESSEX
EAST HAMPSHIRE AONB	HAMPSHIRE *
HIGH WEALD AONB	EAST SUSSEX * KENT * WEST SUSSEX *
ISLE OF WIGHT AONB	ISLE OF WIGHT *
KENT DOWNS AONB	KENT *
NORTH WESSEX DOWNS AONB	BERKSHIRE * HAMPSHIRE * OXFORDSHIRE *
SOUTH HAMPSHIRE COAST AONB	HAMPSHIRE *
SURREY HILLS AONB	SURREY *
SUSSEX DOWNS AONB	EAST SUSSEX * WEST SUSSEX *

19.6 Permissions for sand and gravel workings in the region accounted for 37% of the total area for this mineral permitted in England. Chalk permissions in the region made up 61% of the England total and clay/shale almost 35%. Oil/gas exploration had permitted areas covering less than 80 ha, however, this represented 46% of the England total (Table 19.2).

19.7 Over 70% of surface mineral workings (13060 ha) had planning permissions which included satisfactory reclamation conditions. A further 21% (3770 ha) had unsatisfactory reclamation conditions. Permissions which had no provisions for reclamation accounted for less than 7% of the total (1220 ha), and 2% (430 ha) had provisions for reclamation which were unlikely to be fulfilled (Figure 19.2).

19.8 Sand and gravel and clay/shale permissions had the largest areas with unsatisfactory reclamation conditions covering 1200 ha and 1020 ha respectively (Figure 19.3). For sand and gravel this compared with 9080 ha (84%) with satisfactory conditions whereas clay/shale sites had only 2290 ha (65%) with satisfactory conditions.

19.9 All 640 ha of ironstone permissions (Oxfordshire, 1982 data) had unsatisfactory conditions, although most of this was unworked (620 ha) and is likely to remain so, at least for the original mineral (see paragraph 7.3). Limestone, chalk, and particularly sandstone also had proportionally small areas covered by satisfactory reclamation conditions (46%, 37%, and 7% respectively), although in terms of the total areas involved, chalk was much the most important (Figure 19.3).

TABLE 19.2 TOTAL AREA PERMITTED FOR SURFACE MINERAL WORKINGS IN THE SOUTH EAST REGION, INDICATING THE PERCENTAGE REGIONAL CONTRIBUTION TO THE ENGLAND TOTAL.

MINERAL	TOTAL AREA PERMITTED IN REGION (ha)	TOTAL AREA PERMITTED IN ENGLAND (ha)	REGIONAL CONTRIBUTION TO ENGLAND TOTAL (%)
CHALK	2240	3660	61.2
CLAY/SHALE	3500	10090	34.7
IRONSTONE	640	14420	4.4
LIMESTONE	480	11490	4.2
OIL/GAS EXPLORATION	40	80	46.1
OIL/GAS PRODUCTION	10	60	19.7
SAND & GRAVEL	10780	29040	37.1
INDUSTRIAL SAND	270	2130	12.7
SANDSTONE	50	2940	1.7
OTHER MINERALS	410	6530	6.3
TOTAL	18410		

Note: Areas are rounded to the nearest 10 hectares.

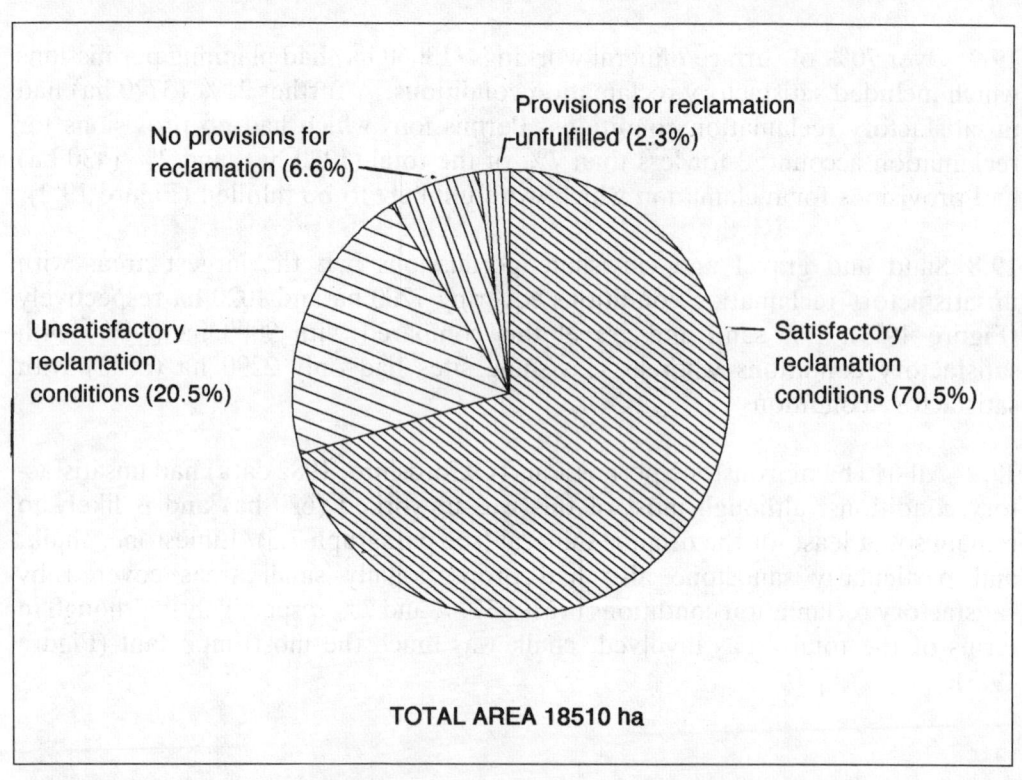

FIGURE 19.2 NATURE OF PROVISIONS FOR RECLAMATION FOR SURFACE MINERAL WORKINGS IN THE SOUTH EAST REGION IN 1988.

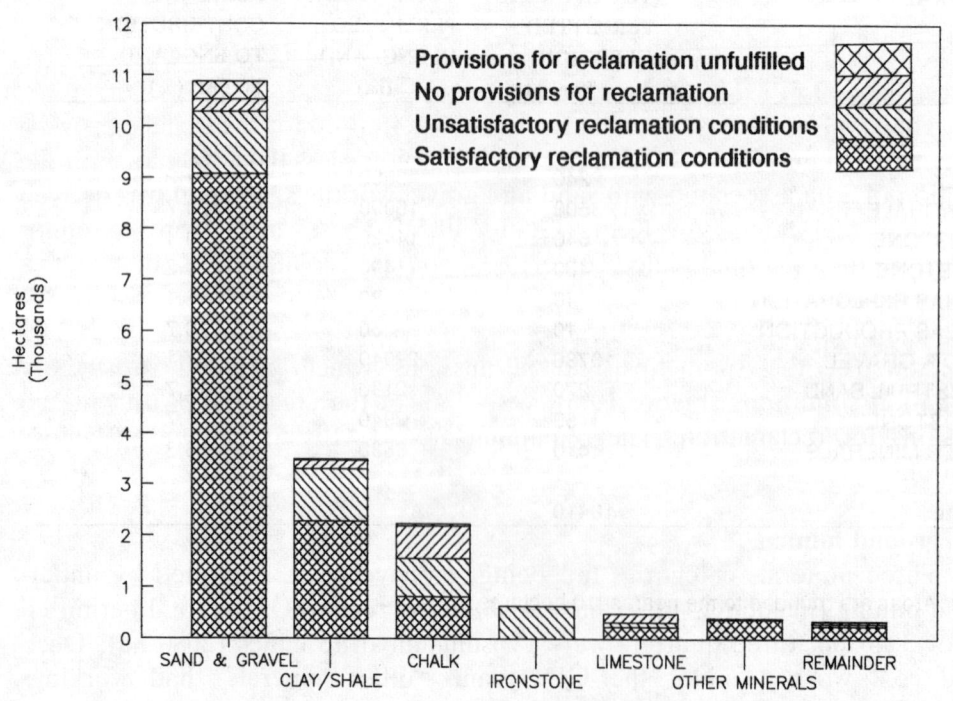

FIGURE 19.3 AREA AND NATURE OF PROVISIONS FOR RECLAMATION OF SURFACE MINERAL WORKINGS IN THE SOUTH EAST REGION IN 1988 – MAIN MINERAL TYPES.

19.10 Over 3500 ha of surface mineral working required imported fill to achieve reclamation - 19% of the total area of surface mineral workings permitted in the region. This was considerably higher than the national average of 10%, and accounts for 37% of the total area of land dependant on fill recorded in England. Over 90% of the area involved sand and gravel or clay/shale sites (1910 ha and 1300 ha respectively) (Figure 19.4).

FIGURE 19.4 SURFACE MINERAL WORKINGS DEPENDENT ON IMPORTED FILL IN THE SOUTH EAST REGION IN 1988 – MAIN MINERAL TYPES

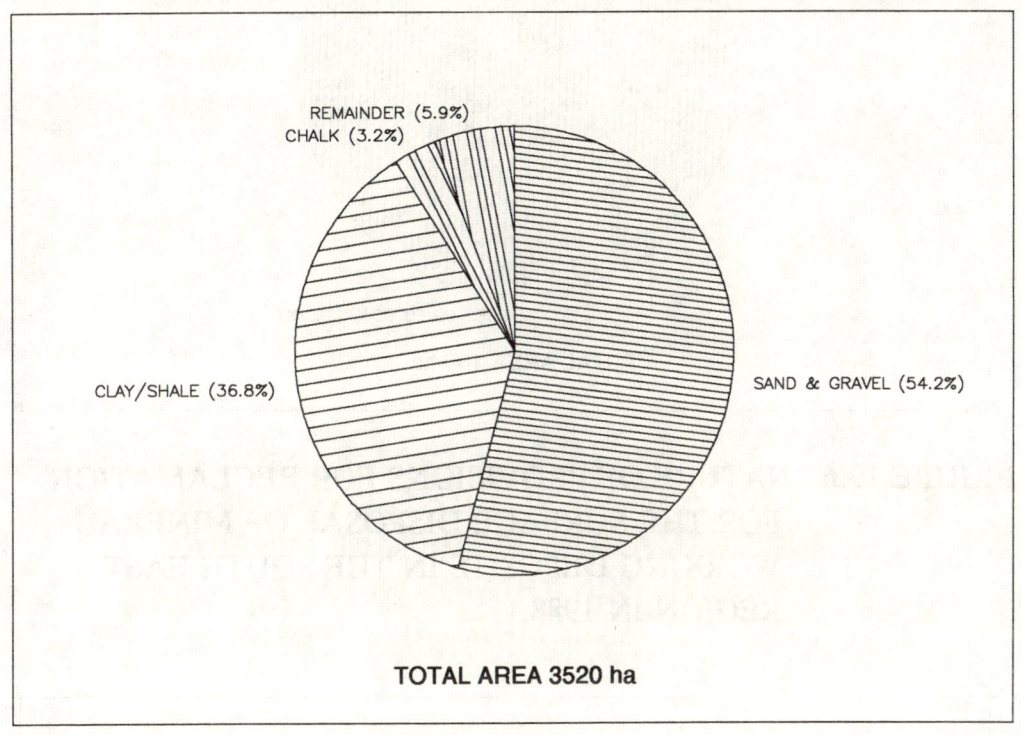

Surface Disposal of Mineral Working Deposits

19.11 There was a relatively small area of land permitted for the surface disposal of mineral working deposits (spoil tips) in the region - less than 160 ha. With the exception of 2 ha associated with sand and gravel workings, all spoil was derived from deep coal mining in Kent (Figure 19.5). These areas were of minor significance on a national scale (Table 19.3), and the last of the 3 Kent collieries has now closed.

19.12 None of the spoil tips had permissions which included satisfactory reclamation conditions (Figure 19.6), with 97% of the area (150 ha) having no provisions for reclamation. The remaining area, (<10 ha) had unsatisfactory conditions.

Underground mining

19.13 Three minerals categories are being, or have been, exploited by underground mining (including "other minerals") in the region. Of the total permitted area of 3640 ha, three quarters were gypsum/anhydrite mines (2680 ha). Deep mined coal worked under the GDO, and "other minerals" had workings extending 940 ha and 20 ha respectively.

FIGURE 19.5 AREA PERMITTED FOR THE SURFACE
            DISPOSAL OF MINERAL WORKING DEPOSITS
            IN THE SOUTH EAST REGION IN 1988
            – MAIN MINERAL TYPES.

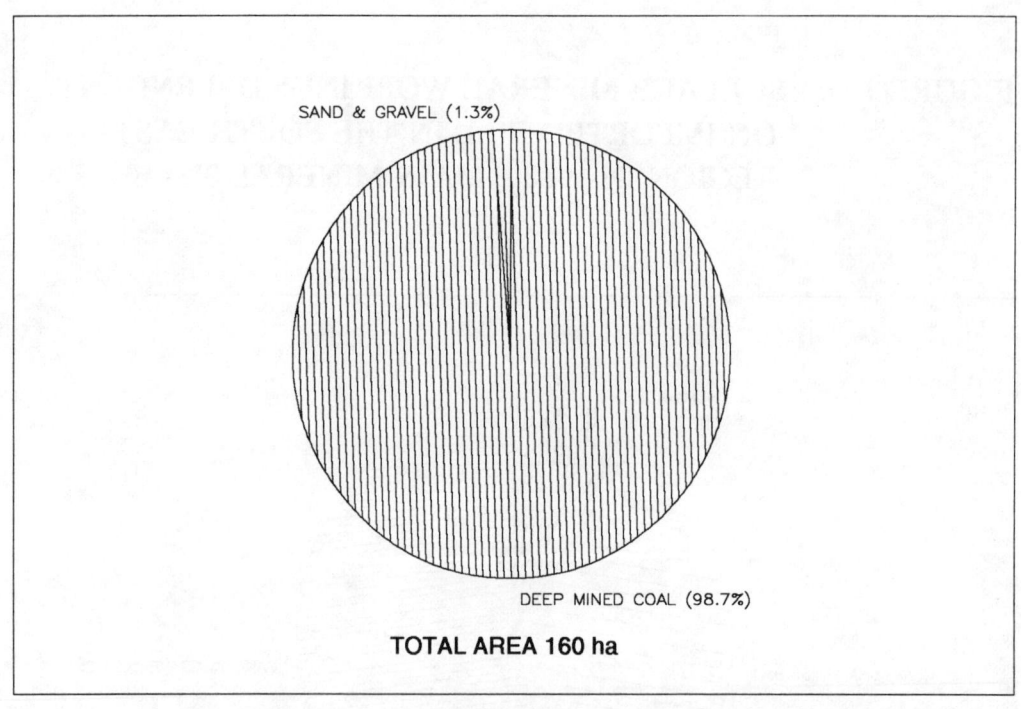

FIGURE 19.6 NATURE OF PROVISIONS FOR RECLAMATION
            FOR THE SURFACE DISPOSAL OF MINERAL
            WORKING DEPOSITS IN THE SOUTH EAST
            REGION IN 1988.

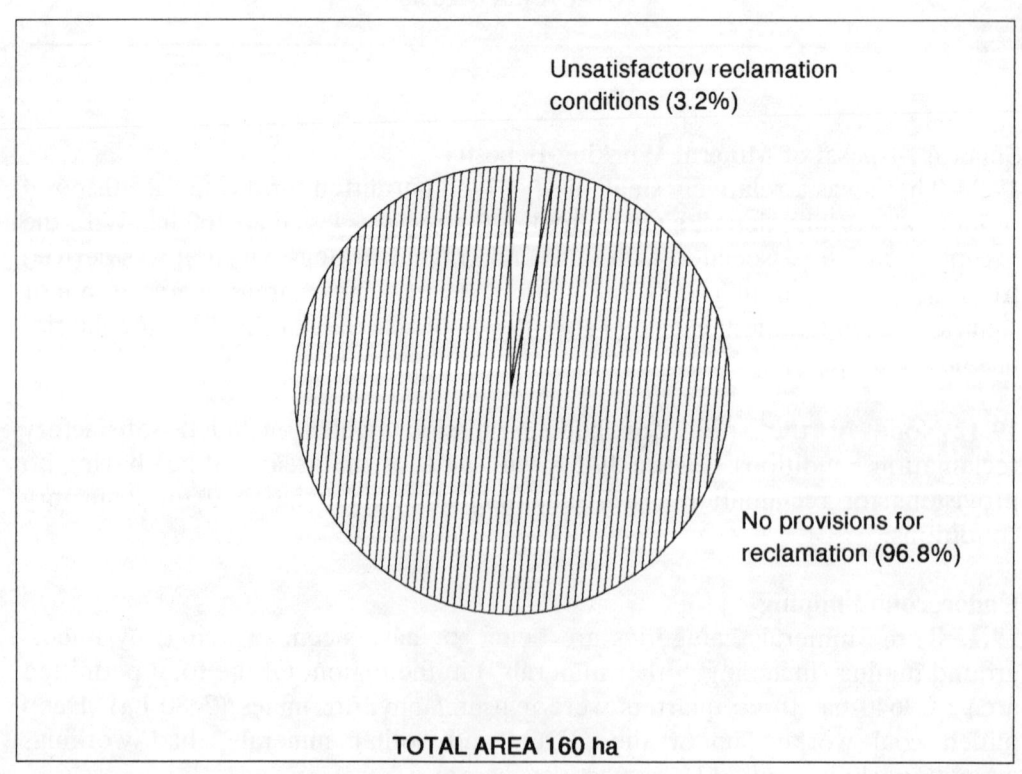

TABLE 19.3 TOTAL AREA PERMITTED FOR THE SURFACE DISPOSAL OF MINERAL WORKING DEPOSITS IN THE SOUTH EAST REGION, INDICATING THE PERCENTAGE REGIONAL CONTRIBUTION TO THE ENGLAND TOTAL.

MINERAL	TOTAL AREA PERMITTED IN REGION (ha)	TOTAL AREA PERMITTED IN ENGLAND (ha)	REGIONAL CONTRIBUTION TO ENGLAND TOTAL (%)
DEEP MINED COAL	150	11120	1.3
SAND & GRAVEL	<10	140	1.4
TOTAL	150		

Note: Areas are rounded to the nearest 10 ha.

TABLE 19.4 TOTAL AREA OF UNDERGROUND PERMISSIONS IN THE SOUTH EAST REGION, INDICATING THE PERCENTAGE REGIONAL CONTRIBUTION TO THE ENGLAND TOTAL.

MINERAL	TOTAL AREA PERMITTED IN REGION (ha)	TOTAL AREA PERMITTED IN ENGLAND (ha)	REGIONAL CONTRIBUTION TO ENGLAND TOTAL (%)
COAL (UNDER GDO)	940	315820	0.3
GYPSUM/ANHYDRITE	2680	14670	18.3
OTHER MINERALS	20	17850	<0.1
TOTAL	3640		

Note: Areas are rounded to the nearest 10 ha.

19.14 The gypsum/anhydrite permissions in the region accounted for over 18% of the total permitted area for underground mining of the mineral in England (Table 19.4).

Land covered by permissions which include aftercare conditions
19.15 There were 222 permissions within the region which included aftercare conditions, the highest number for any region. All conditions were for surface mineral workings and covered an area of 3970 ha. This was equivalent to 22% of the current permitted area in the region.

19.16 Most of the land with permissions which included aftercare conditions (75% or 2960 ha) were sand and gravel workings. Clay/shale and chalk permissions accounted for a further 13% (530 ha) and 6% (230 ha) respectively (Figure 19.7).

FIGURE 19.7 AREA OF MINERAL WORKINGS COVERED BY AFTERCARE CONDITIONS IN THE SOUTH EAST REGION IN 1988 – MAIN MINERAL TYPES, AND PROPOSED ENDUSES.

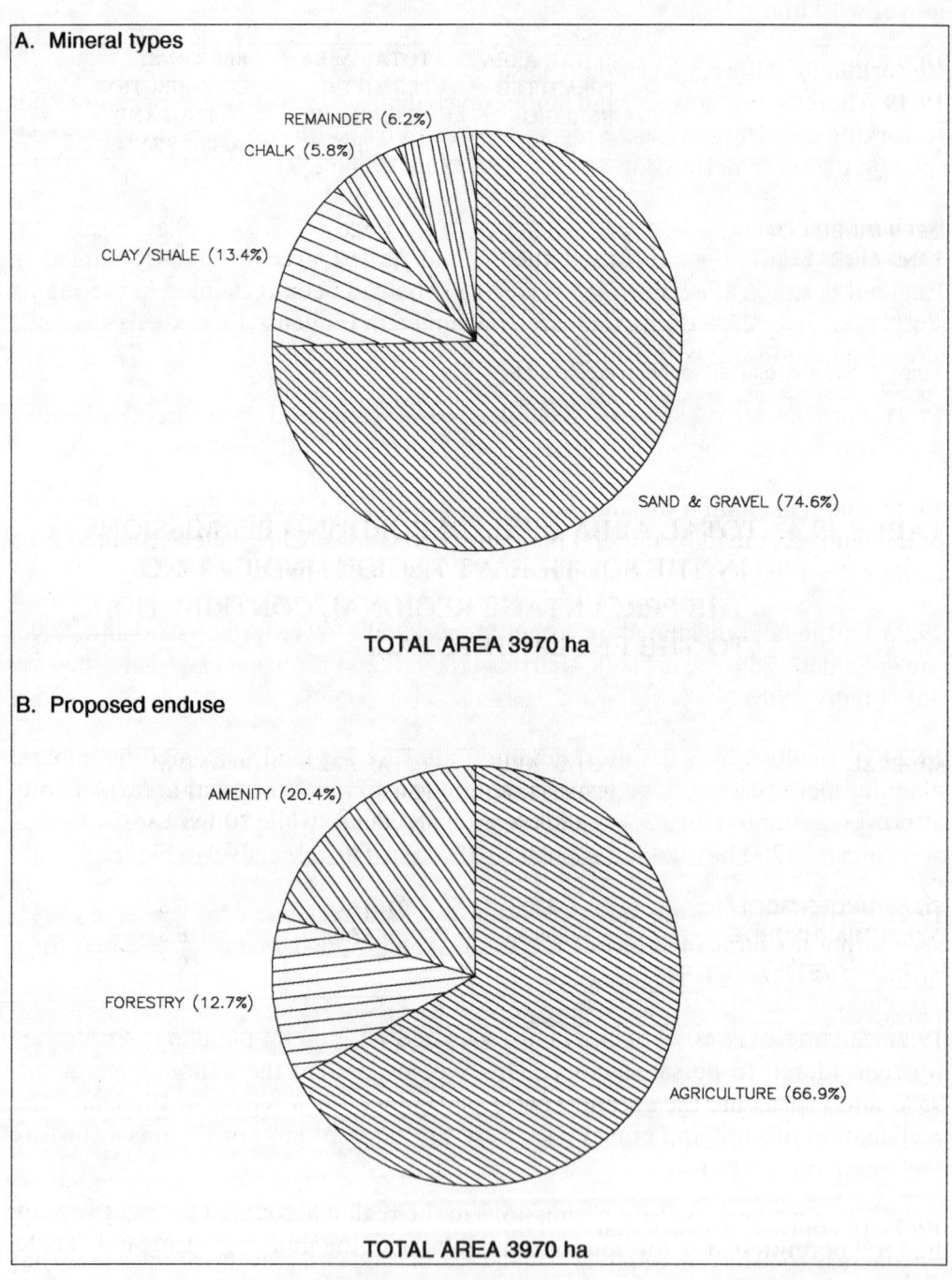

19.17 Agriculture was the most common afteruse proposed, accounting for 67% of the total (2650 ha). Amenity reclamation was proposed for a further 20% (810 ha) and forestry 13% (500 ha) (Figure 19.7).

19.18 Agriculture was recorded as the most common afteruse proposed for all mineral types recorded. Almost all proposals for reclamation to amenity uses (95% or 770 ha) and 83% (420 ha) of reclamation to forestry were on sand and gravel workings.

Reworking of Mineral working deposits
19.19 There were only 2 sand and gravel deposits which were permitted for reworking in the region, covering an area of 13 ha. Both sites were worked under specific planning permissions (see Table J2 in Volume 2).

Reclamation
19.20 The South East recorded the second largest area of land reclaimed in England between 1982 and 1988. Over 4500 ha had been reclaimed in the region since 1982, over 22% of the total area reclaimed in England (these figures exclude any reclamation in Oxfordshire).

19.21 The majority (85% or 3890 ha) were reclaimed by mineral planning permissions, whilst 15% (660 ha) were reclaimed by "other means".

19.22 The reclamation of sand and gravel workings (3150 ha) accounted for 81% of the land reclaimed by mineral planning permissions. Clay/shale accounted for a further 8% (300 ha) and chalk almost 5% (180 ha) (Figure 19.8).

19.23 Of the land reclaimed by "other means", 60% were former sand and gravel sites (400 ha), 28% were chalk quarries (180 ha) and 9% were clay/shale sites (60 ha) (Figure 19.8).

19.24 Agriculture was the most common afteruse for land reclaimed by mineral planning permissions (2210 ha), accounting for 57% of the total area. Amenity afteruses accounted for 27% of reclamation (1030 ha), while "other uses" (390 ha) and forestry (260 ha), added a further 10% and 7% respectively (Figure 19.9).

19.25 "Other uses" were most common for land reclaimed by "other means", accounting for 66% (430 ha) of the total. Amenity reclamation accounted for a further 24% (160 ha) (Figure 19.9).

19.26 Slightly over 90% of the land reclaimed by mineral planning permissions was considered to be satisfactory. This was lower than the national average of 95% and represents the second lowest percentage for a region in England. The reclamation of sand and gravel sites were the cause of 86% of the unsatisfactory reclamation (320 ha).

19.27 In contrast, 97% of land reclaimed by "other means" was considered to be satisfactory (640 ha). A relatively small area of sand and gravel workings (<20 ha) was considered to be unsatisfactory.

Number of Sites
19.28 There were more surface mineral workings in the South East than in any other region. A total of 926 sites were recorded, 21% of all surface workings permitted in England. There were only 4 sites permitted for the surface disposal of mineral working deposits and 5 underground mines.

19.29 Sand and gravel workings were the most numerous (586 sites) accounting for 63% of the total. Chalk accounted for a further 14% (130 sites), and clay/shale 11% (98 sites) (Table 19.5).

FIGURE 19.8 AREA OF LAND RECLAIMED BY MINERAL PLANNING PERMISSIONS (A) AND BY OTHER MEANS (B) IN THE SOUTH EAST REGION BETWEEN 1982 AND 1988 – MAIN MINERAL TYPES.

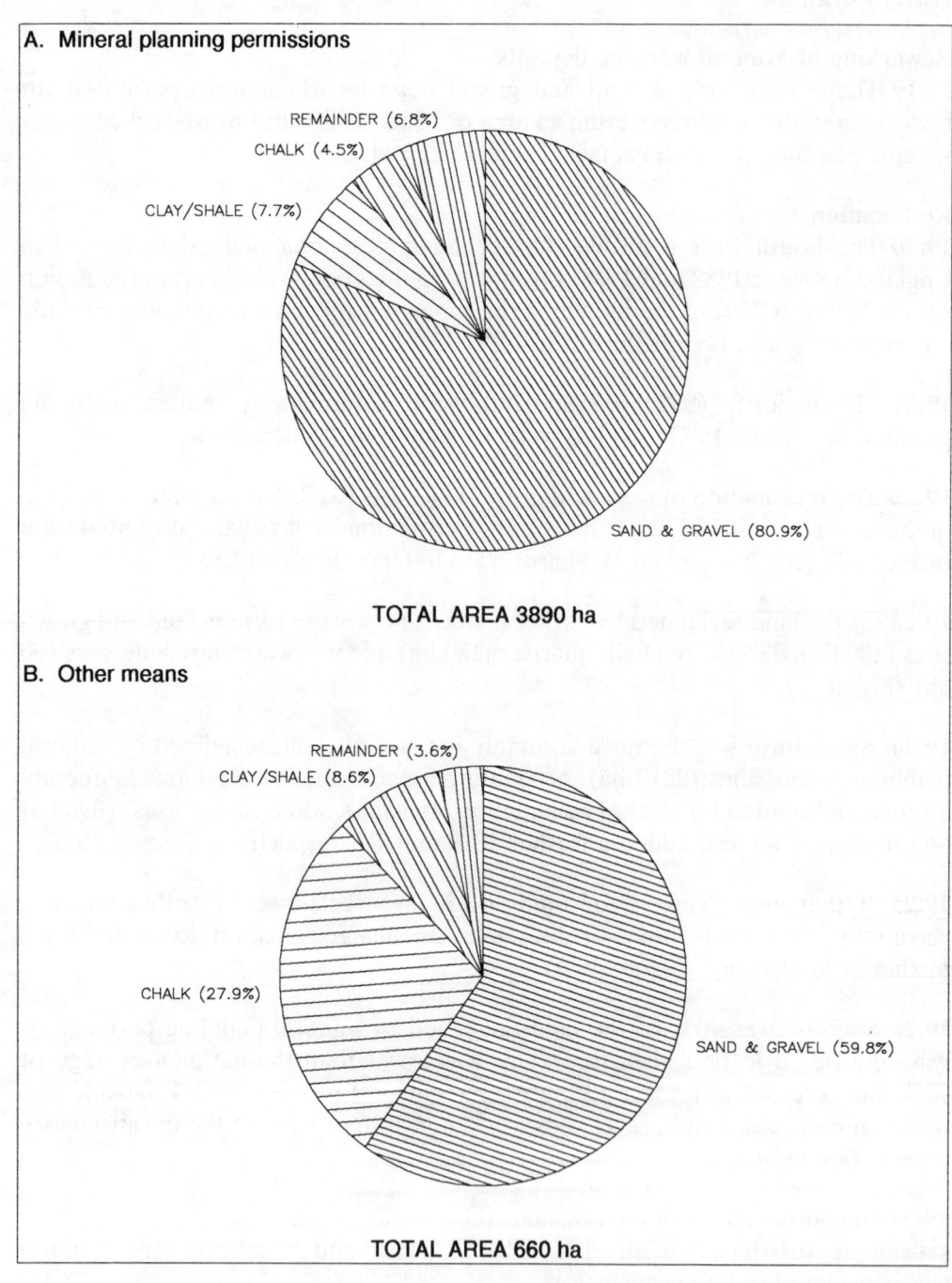

FIGURE 19.9 CHOSEN AFTERUSE OF LAND RECLAIMED BY MINERAL PLANNING PERMISSIONS (A) AND BY OTHER MEANS (B) IN THE SOUTH EAST REGION BETWEEN 1982 AND 1988.

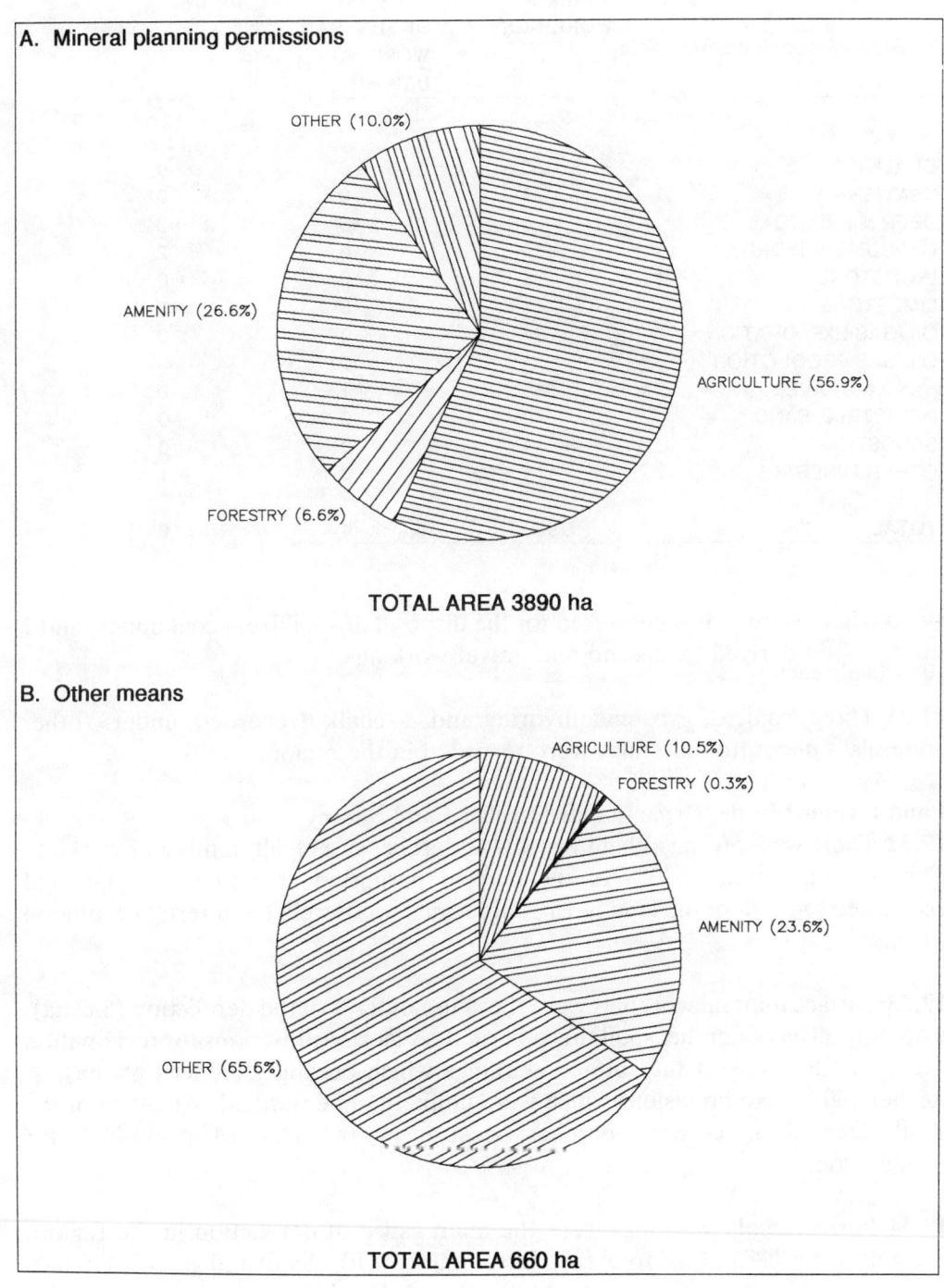

TABLE 19.5 NUMBER OF SITES FOR MINERAL WORKINGS IN THE SOUTH EAST REGION IN 1988.

MINERAL	SURFACE MINERAL WORKINGS	SURFACE DISPOSAL OF MINERAL WORKING DEPOSITS	UNDERGROUND MINING
CHALK	130	0	0
CLAY/SHALE	98	0	0
DEEP MINED COAL	0	3	3
GYPSUM/ANHYDRITE	0	0	2
IRONSTONE	5	0	0
LIMESTONE	41	0	0
OIL/GAS EXPLORATION	34	0	0
OIL/GAS PRODUCTION	1	0	0
SAND & GRAVEL	586	1	0
INDUSTRIAL SAND	12	0	0
SANDSTONE	4	0	0
OTHER MINERALS	15	0	1
TOTAL	926	4	6

19.30 There were 3 sites permitted for the disposal of spoil from coal mines, and 1 site for spoil derived from sand and gravel workings.

19.31 Three coal, 2 gypsum/anhydrite and 1 chalk (recorded under "other minerals") underground mines were recorded in the region.

Land included in the Departments Derelict Land Survey
19.32 There were 560 ha of land which was derelict as a result of mineral working. This was equivalent to 3% of the current permitted area in the region and accounted for 10% of the land in England which was derelict as a result of mineral working.

19.33 Surface mineral workings were the cause of 95% of the dereliction (530 ha), with only 30 ha of derelict spoil tips recorded. A lack of provisions for reclamation accounted for 74% of the surface workings which became derelict (390 ha). A further 140 ha had provisions which were unlikely to be fulfilled. All 30 ha of the spoil disposal areas were derelict because of the lack of provisions for reclamation.

19.34 Former chalk workings were the main cause of dereliction in the region, accounting for 68% of the total (380 ha) (Figure 19.10). Sand and gravel (110 ha), spoil derived from deep mined coal (30 ha) and clay/shale (30 ha) sites were also recorded as derelict.

Reviews under Section 264A of the Town and Country Planning Act
19.35 Four mineral planning authorities had started their mineral reviews; Hampshire, Isle of Wight, Kent and Surrey.

19.36 The reviews had covered 266 sites and 4 unimplemented permissions Of these 72 had been identified for further action (27%), although no action had been identified.

19.37 The sites identified for further action were 49 sand and gravel workings, 12 chalk, 5 clay/shale, 5 limestone/dolomite and 1 "other minerals" site.

FIGURE 19.10 AREA OF LAND DAMAGED BY MINERAL WORKINGS IN THE SOUTH EAST AND INCLUDED IN THE 1988 DERELICT LAND SURVEY

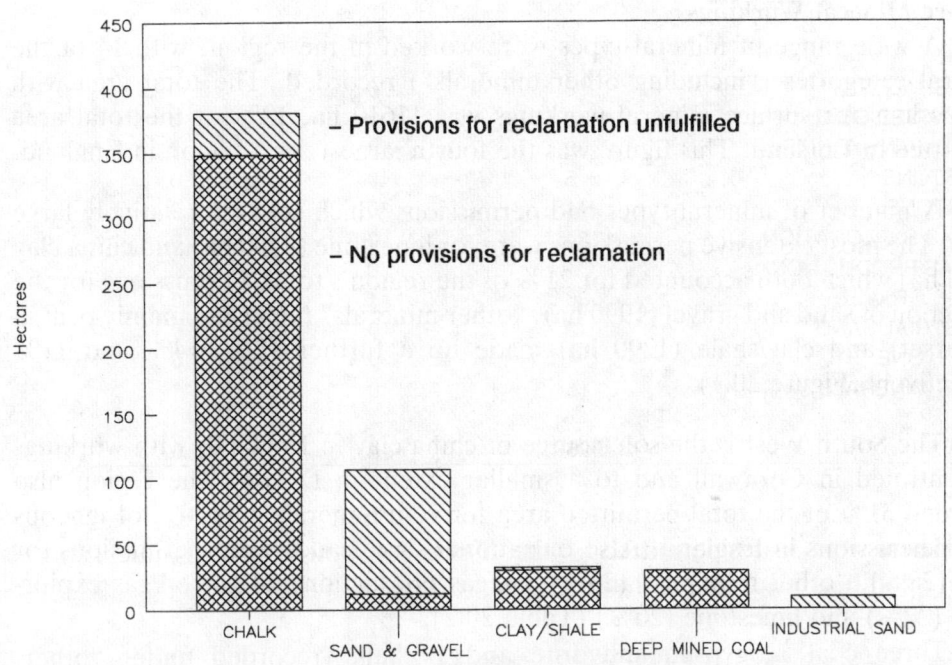

20. SOUTH WEST

Introduction

20.1 The South West region is made up of 7 mineral planning authorities; Avon, Cornwall, Devon, Dorset, Gloucestershire, Somerset and Wiltshire. Mineral workings were recorded by all mineral planning authorities'.

20.2 There are two National Parks and twelve AONB's wholly or partly in the region. Minerals are worked in one of the National Parks (Dartmoor NP), and eleven of the AONB's as indicated in Table 20.1.

Surface Mineral Workings

20.3 A wide range of mineral types were worked in the region, with 14 of the mineral categories ("including other minerals") recorded. The total area with permissions for surface mineral workings was 11610 ha, 12% of the total area permitted in England. This figure was the fourth largest for a region in England.

20.4 A number of mineral types had permissions which covered relatively large areas. The most extensive permissions were for limestone (2490 ha) and china clay (2380 ha) which both accounted for 21% of the region's total. Permissions for the extraction of sand and gravel (1990 ha), "other minerals" (1630 ha - mainly peat in Somerset) and clay/shale (1290 ha) made up a further 17%, 14% and 11% respectively (Figure 20.1).

20.5 The South West is the sole source of china clay in England, with workings concentrated in Cornwall and to a smaller extent in Devon. The region also recorded 31% of the total permitted area for vein minerals and 30% of igneous rock permissions in England. Also of national significance were permissions for slate (28%), "other minerals" (25%), oil/gas production (24%), oil/gas exploration (22%) and limestone (20%) (Table 20.2).

FIGURE 20.1 AREA OF SURFACE MINERAL WORKINGS PERMITTED IN THE SOUTH WEST REGION IN 1988 – MAIN MINERAL TYPES.

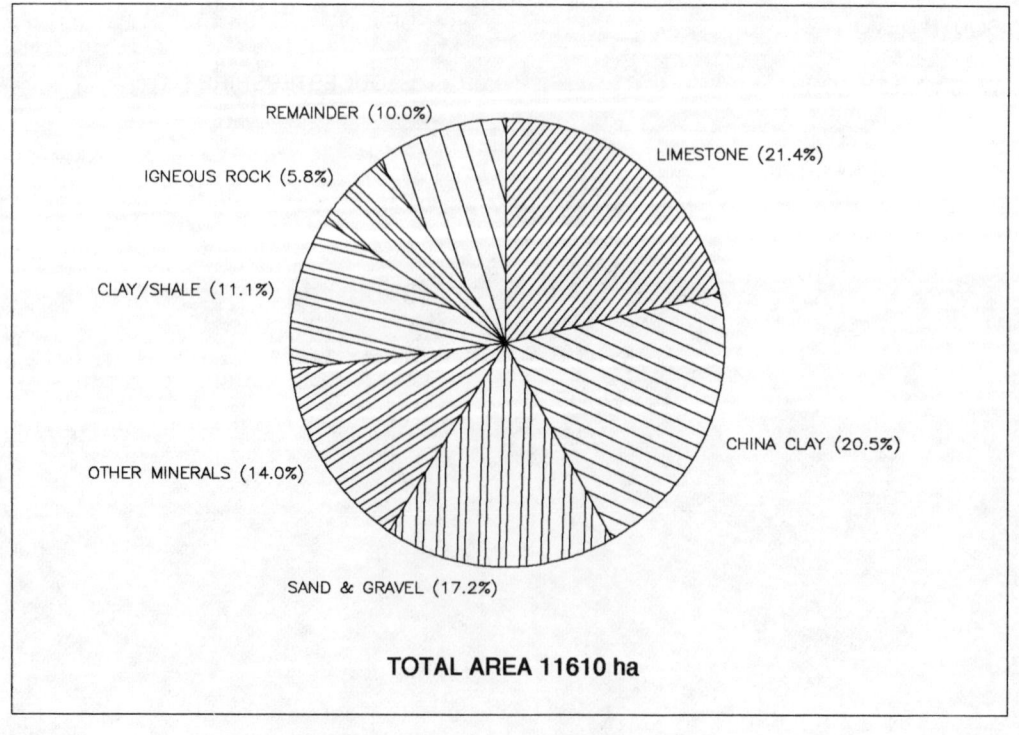

TABLE 20.1 NATIONAL PARKS AND AREAS OF OUTSTANDING NATURAL BEAUTY IN THE SOUTH WEST REGION, INDICATING RELEVANT MINERAL PLANNING AUTHORITIES AND THOSE WITH MINERAL WORKINGS (*).

NP or AONB	MPA
CORNWALL AONB	CORNWALL *
COTSWOLDS AONB	AVON *
	GLOUCESTERSHIRE *
	WILTSHIRE *
CRANBOURNE CHASE AND WEST WILTSHIRE DOWNS AONB	DORSET
	WILTSHIRE *
DARTMOOR NP	DEVON *
DORSET AONB	DORSET *
EAST DEVON AONB	DEVON *
EXMOOR NP	DEVON
	SOMERSET
MALVERN HILLS AONB	GLOUCESTERSHIRE
MENDIP HILLS AONB	AVON *
	SOMERSET *
NORTH DEVON AONB	DEVON *
NORTH WESSEX DOWNS AONB	WILTSHIRE *
QUANTOCK HILLS AONB	SOMERSET *
SOUTH DEVON AONB	DEVON *
WYE VALLEY AONB	GLOUCESTERSHIRE *

TABLE 20.2 TOTAL AREA PERMITTED FOR SURFACE MINERAL WORKINGS IN THE SOUTH WEST REGION, INDICATING THE PERCENTAGE REGIONAL CONTRIBUTION TO THE ENGLAND TOTAL.

MINERAL	TOTAL AREA PERMITTED IN REGION (ha)	TOTAL AREA PERMITTED IN ENGLAND (ha)	REGIONAL CONTRIBUTION TO ENGLAND TOTAL (%)
CHALK	140	3660	3.8
CHINA CLAY	2380	2380	100.0
CLAY/SHALE	1290	10090	12.8
OPENCAST COAL	40	8420	0.5
IGNEOUS ROCK	670	2240	29.9
LIMESTONE	2490	11490	21.7
OIL/GAS EXPLORATION	20	80	22.4
OIL/GAS PRODUCTION	20	60	24.6
SAND & GRAVEL	1990	29040	6.9
INDUSTRIAL SAND	80	2130	3.6
SANDSTONE	310	2940	10.9
SLATE	90	310	28.3
VEIN MINERALS	480	1540	31.2
OTHER MINERALS	1630	6530	25.0
TOTAL	11610		

Note: Areas are rounded to the nearest 10 hectares.

20.6 A relatively small percentage of the area permitted for surface workings had satisfactory reclamation conditions (27% or 3140 ha). This was the worst position for any region. Almost 23% of the permitted area (2610 ha) had unsatisfactory reclamation conditions and half (5790 ha) had no provisions for reclamation. Land where provisions for reclamation were unlikely to be fulfilled accounted for less than 1% of the total area (70 ha) (Figure 20.2).

20.7 The region's poor record for adequate provisions for reclamation was shared amongst all the most extensively worked minerals (Figure 20.3). Limestone, the most widespread mineral type had only 16% of the permitted area (390 ha) covered by satisfactory reclamation conditions. Almost 1600 ha (64%) had no provisions for reclamation, and 500 ha (20%) had unsatisfactory reclamation conditions. Similarly only 32% of china clay permissions (760 ha), 20% of clay/shale (260 ha), 19% of igneous rock (120 ha) and 15% of "other minerals" (240 ha) had satisfactory reclamation conditions.

20.8 Only 135 ha required imported fill to achieve reclamation, 1% of the permitted area. Over half of this area (52% or 70 ha) were sand and gravel workings (Figure 20.4). Small areas of clay/shale, limestone and industrial sand sites made up most of the remainder (50 ha in total).

Surface Disposal of Mineral Working Deposits
20.9 Eleven mineral types were recorded with permissions for the surface disposal of mineral working deposits (ie spoil tips), covering a total area of 5260 ha (Table 20.3). This was the largest area permitted for spoil tips for any region in England and accounted for 29% of the England figure.

FIGURE 20.2 NATURE OF PROVISIONS FOR RECLAMATION FOR SURFACE MINERAL WORKINGS IN THE SOUTH WEST REGION IN 1988.

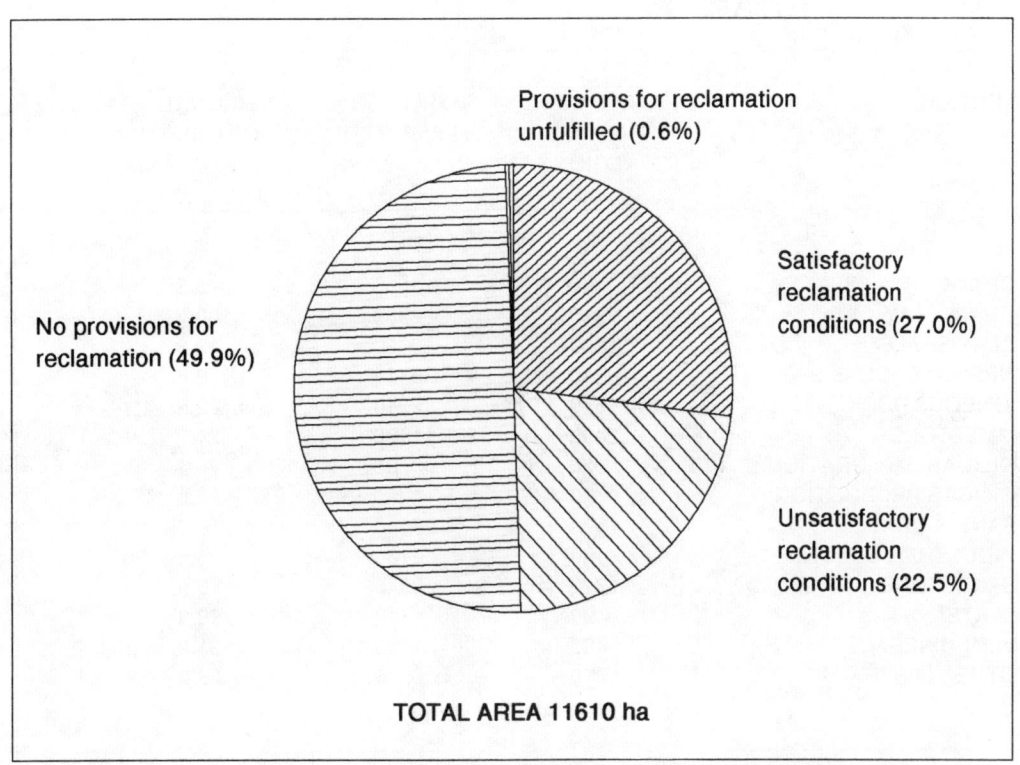

FIGURE 20.3 AREA AND NATURE OF PROVISIONS FOR RECLAMATION OF SURFACE MINERAL WORKINGS IN THE SOUTH WEST REGION IN 1988 – MAIN MINERAL TYPES.

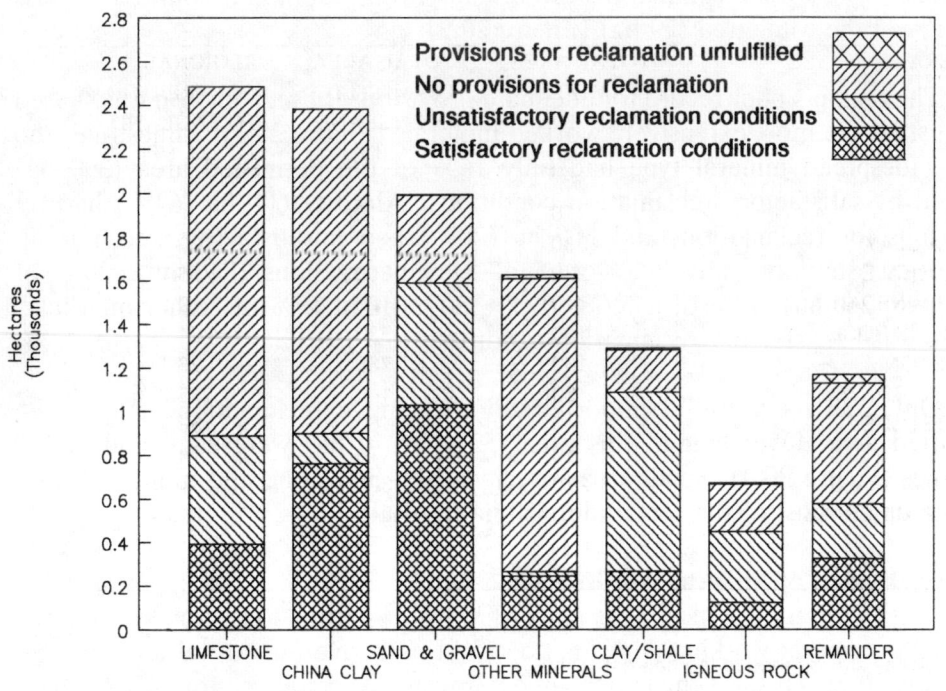

FIGURE 20.4 SURFACE MINERAL WORKINGS DEPENDENT ON IMPORTED FILL IN THE SOUTH WEST REGION IN 1988 – MAIN MINERAL TYPES

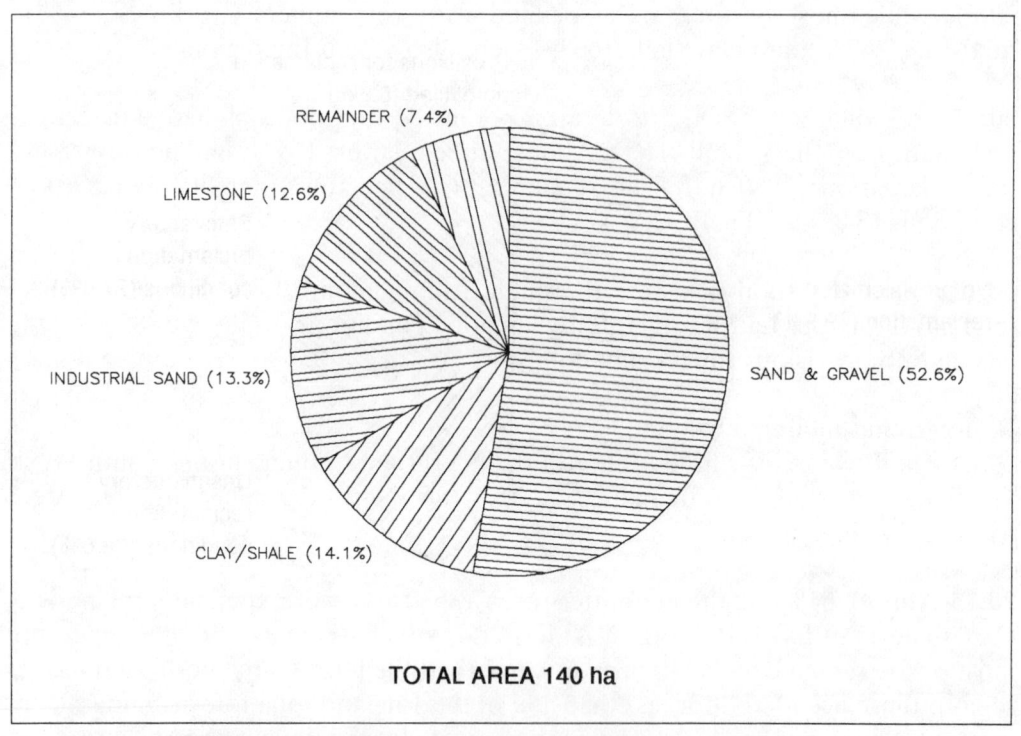

TABLE 20.3 TOTAL AREA PERMITTED FOR THE SURFACE DISPOSAL OF MINERAL WORKING DEPOSITS IN THE SOUTH WEST REGION, INDICATING THE PERCENTAGE REGIONAL CONTRIBUTION TO THE ENGLAND TOTAL.

MINERAL	TOTAL AREA PERMITTED IN REGION (ha)	TOTAL AREA PERMITTED IN ENGLAND (ha)	REGIONAL CONTRIBUTION TO ENGLAND TOTAL (%)
CHALK	<10	<10	–
CHINA CLAY	4330	4330	100.0
CLAY/SHALE	230	350	65.7
DEEP MINED COAL	<10	11120	<0.1
IGNEOUS ROCK	130	530	24.5
LIMESTONE	180	780	23.1
SAND & GRAVEL	50	140	35.7
SANDSTONE	40	60	66.7
SLATE	10	70	14.3
VEIN MINERALS	270	320	84.4
OTHER MINERALS	10	40	25.0
TOTAL	5260		

Note: Areas are rounded to the nearest 10 ha.

20.10 Wastes derived from china clay workings, accounted for 82% of the total permitted area in the region (4330 ha) (Figure 20.5). Spoil tips associated with the extraction of vein minerals (270 ha), clay/shale (230 ha), limestone (180 ha) and igneous rock (130 ha) accounted for most of the remainder.

20.11 All china clay spoil tips in England were in the South West region (Table 20.3), while most of the tips associated with vein mineral workings (84%), sandstone (67%) and clay/shale (66%) were also within the region.

20.12 Less than 20% of the total area permitted for spoil tips had satisfactory reclamation conditions (1040 ha). Almost three quarters (74%) had no provisions for reclamation (3880 ha). The remaining 6% had unsatisfactory reclamation conditions (340 ha) (Figure 20.6).

20.13 China clay spoil tips were the main planning control problem. Only 13% (560 ha) of the tips had satisfactory reclamation conditions, while over 85% (3700 ha) had no provisions for reclamation.

Underground mining

20.14 There were 4380 ha permitted for underground mining in the South West involving 5 mineral types (Table 20.4). This was the third smallest area recorded for a region in England.

20.15 Almost 81% of the permitted area (3530 ha) were vein mineral mines. Clay/shale (580 ha), limestone (170 ha) and "other minerals" (90 ha) made up most of the remainder. While vein minerals had the largest area permitted in the region, these accounted for less than 1% of the England total for this mineral. In contrast, limestone and clay/shale permissions, while involving much smaller areas, made up 57% and 40% of the national totals for these minerals (Table 20.4).

TABLE 20.4 TOTAL AREA OF UNDERGROUND PERMISSIONS IN THE SOUTH WEST REGION, INDICATING THE PERCENTAGE REGIONAL CONTRIBUTION TO THE ENGLAND TOTAL.

MINERAL	TOTAL AREA PERMITTED IN REGION (ha)	TOTAL AREA PERMITTED IN ENGLAND (ha)	REGIONAL CONTRIBUTION TO ENGLAND TOTAL (%)
CLAY/SHALE	580	1470	39.5
COAL (SPEC PLAN PERM)	<10	46380	0.0
LIMESTONE	170	300	56.7
VEIN MINERAL	3530	379710	0.9
OTHER MINERALS	90	17850	0.5
TOTAL	4380		

Note: Areas are rounded to the nearest 10 ha.

Land covered by permissions which include aftercare conditions

20.16 A total of 86 permissions included aftercare conditions, covering an area of 1560 ha. Surface mineral workings accounted for 77 of the permissions and involved 950 ha (61% of the area). There were 9 permissions for spoil tips with

FIGURE 20.5 AREA PERMITTED FOR THE SURFACE DISPOSAL OF MINERAL WORKING DEPOSITS IN THE SOUTH WEST REGION IN 1988 – MAIN MINERAL TYPES.

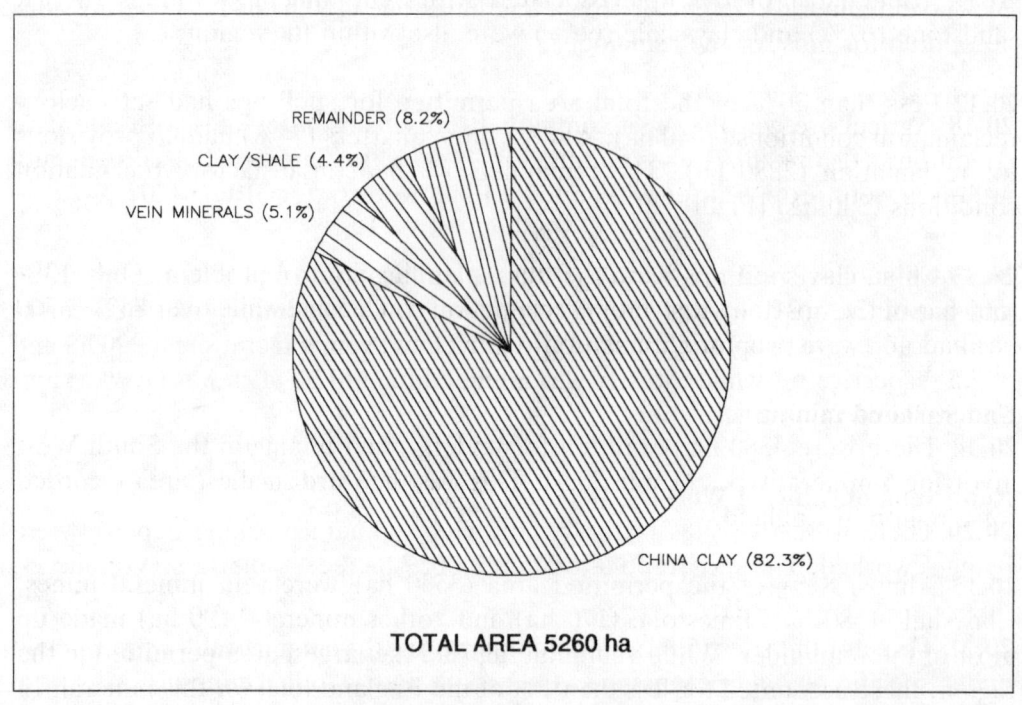

FIGURE 20.6 NATURE OF PROVISIONS FOR RECLAMATION FOR THE SURFACE DISPOSAL OF MINERAL WORKING DEPOSITS IN THE SOUTH WEST REGION IN 1988.

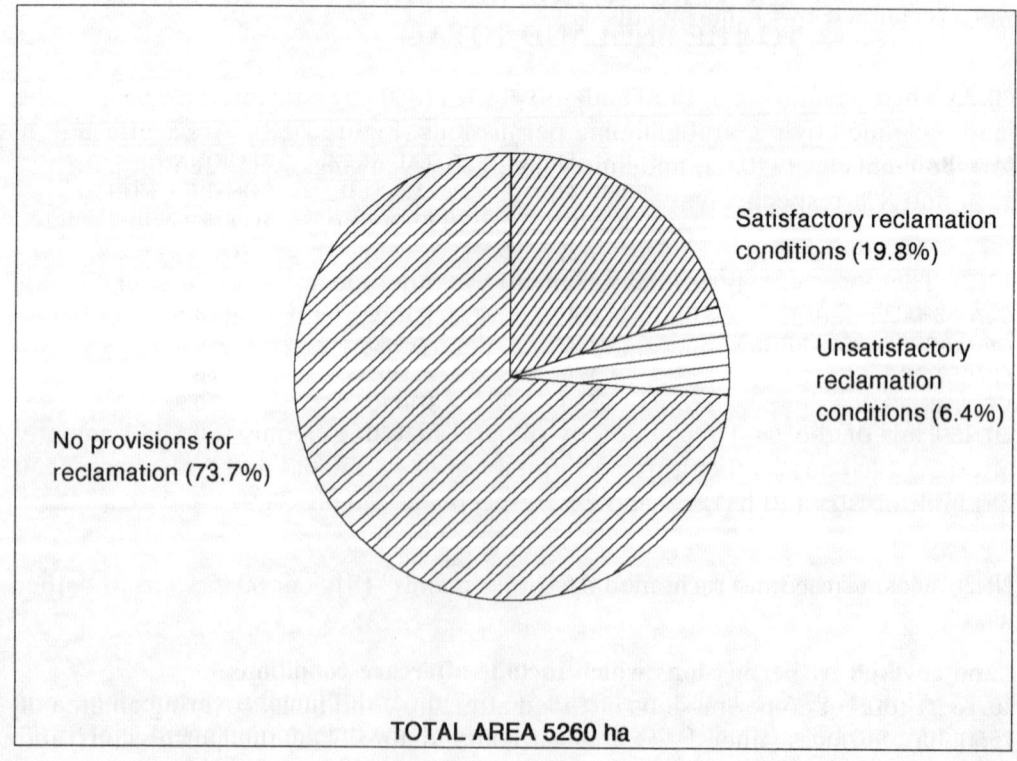

aftercare conditions covering an area of 620 ha. These figures represent 8% of the current permitted area of surface workings and 12% of spoil tips. These were both lower than the national averages (18% and 14% respectively).

20.17 Sand and gravel workings with aftercare conditions were the most extensive (550 ha), accounting for 35% of the total (Figure 20.7). China clay (350 ha or 24%), vein minerals (250 ha or 16%), and limestone (160 ha or 10%) permissions accounted for most of the remainder.

20.18 Agriculture was the most common afteruse chosen for sites with aftercare conditions (Figure 20.7), accounting for 43% of the area (670 ha). Amenity afteruses were proposed for 580 ha (37%) and forestry for 310 ha (20%).

20.19 Most of the reclamation to agriculture was proposed for china clay and sand and gravel sites (210 ha and 160 ha respectively). Considerable areas of amenity reclamation were proposed for sand and gravel and vein mineral sites (390 ha and 80 ha respectively), while forestry was intended for 140 ha of china clay workings and 80 ha of vein mineral sites.

Reworking of Mineral Working Deposits
20.20 There were five sites in the region where mineral working deposits were being reworked. These covered a total area of 37 ha (see Table J2 in Volume 2).

20.21 Three of the sites (2 limestone/dolomite and 1 vein minerals) were worked under specific planing permissions. One chalk and 1 "other minerals" site were worked under the GDO.

Reclamation
20.22 The South West region recorded the smallest area of mineral workings reclaimed between 1982 and 1988 in England. Less than 1000 ha were reclaimed, 5% of reclamation in England over that period. Reclamation by mineral planning permissions accounted for 92% of the reclamation (920 ha), while 8% (70 ha) were reclaimed by "other means".

20.23 The reclamation of sand and gravel sites (580 ha) accounted for 63% of the land reclaimed by mineral planning permissions (Figure 20.8). Areas affected by former china clay (110 ha) and limestone (70 ha) workings accounted for a further 12% and 8% respectively.

20.24 There was very little land reclaimed by "other means" in the region (70 ha), and only 2 mineral types had notable areas reclaimed by this method - clay/shale (30 ha) and limestone/ dolomite (30 ha) (Figure 20.8).

20.25 Half of the land reclaimed by mineral planning permissions had amenity afteruses (460 ha). Agriculture accounted for 40% (370 ha) and "other uses" (70 ha) and forestry (20 ha) 8% and 2% respectively (Figure 20.9).

20.26 Most of the land reclaimed by "other means" (81% or 60 ha) was to "other uses".

20.27 Almost 92% of the reclamation by mineral planning permissions was considered to be satisfactory (850 ha), slightly below the national average of 95%. All land reclaimed by "other means" was satisfactory.

FIGURE 20.7 AREA OF MINERAL WORKINGS COVERED BY AFTERCARE CONDITIONS IN THE SOUTH WEST REGION IN 1988 – MAIN MINERAL TYPES, AND PROPOSED ENDUSES.

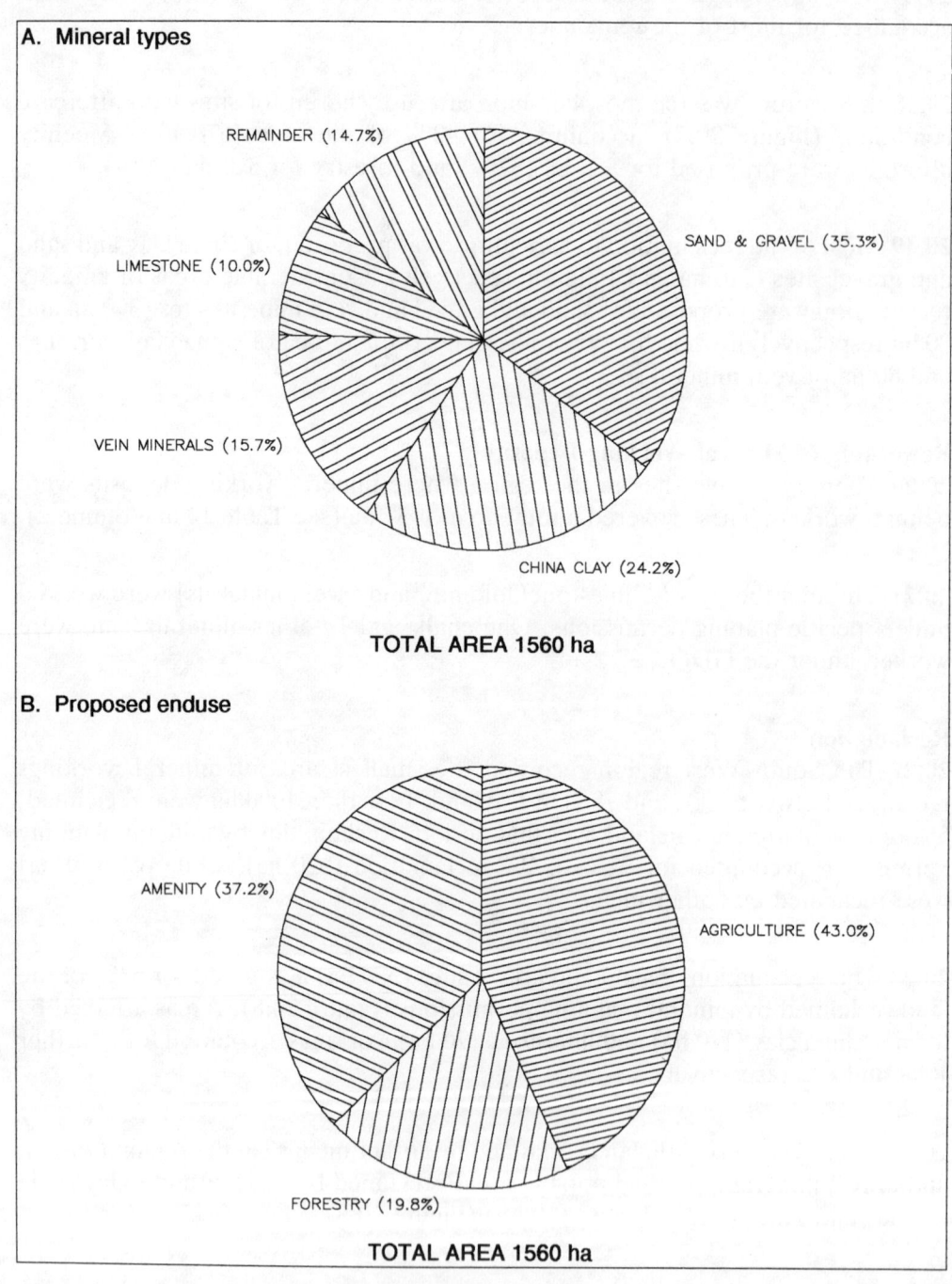

FIGURE 20.8 AREA OF LAND RECLAIMED BY MINERAL PLANNING PERMISSIONS (A) AND BY OTHER MEANS (B) IN THE SOUTH WEST REGION BETWEEN 1982 AND 1988 – MAIN MINERAL TYPES.

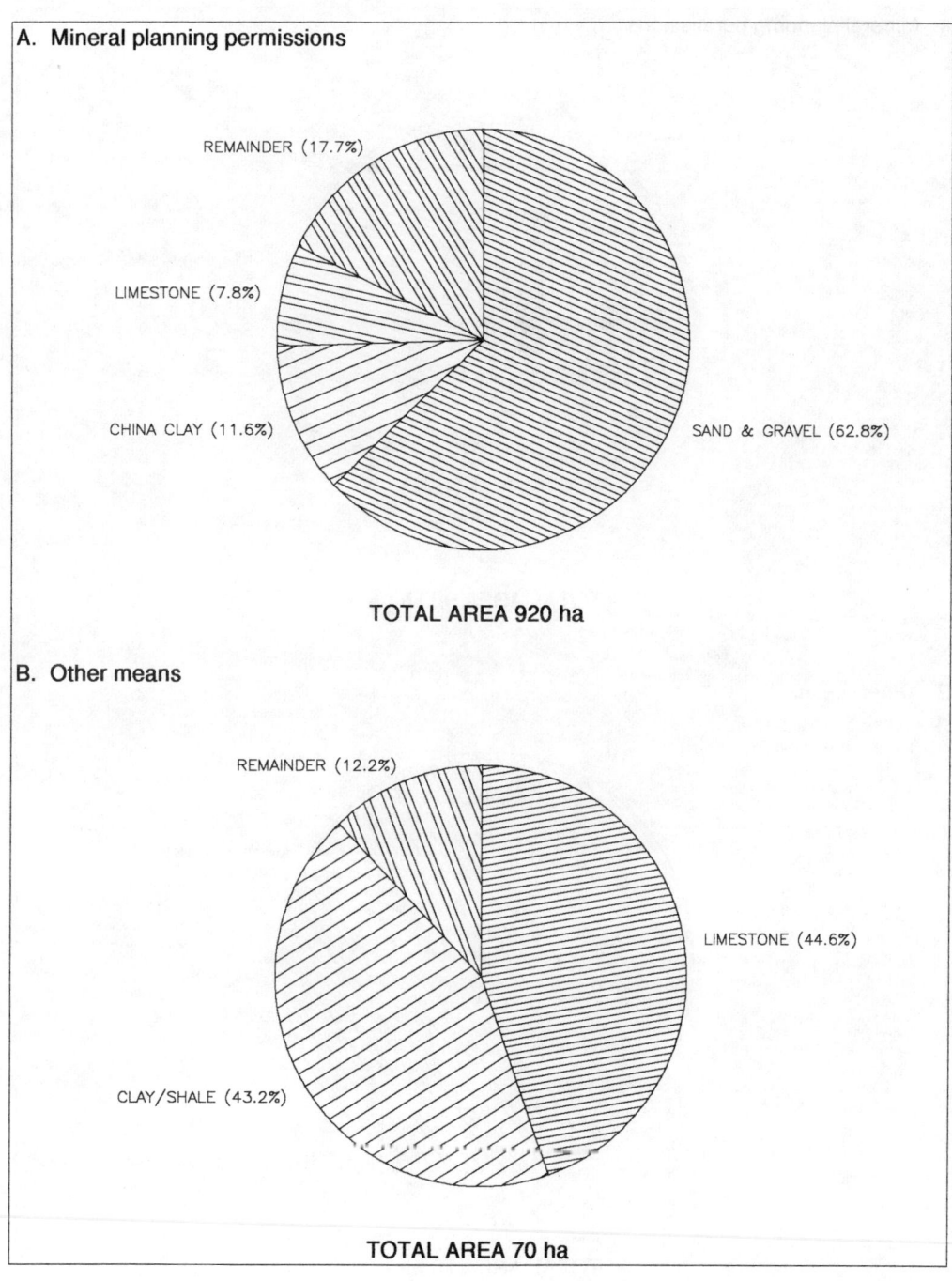

FIGURE 20.9 CHOSEN AFTERUSE OF LAND RECLAIMED BY MINERAL PLANNING PERMISSIONS (A) AND BY OTHER MEANS (B) IN THE SOUTH WEST REGION BETWEEN 1982 AND 1988.

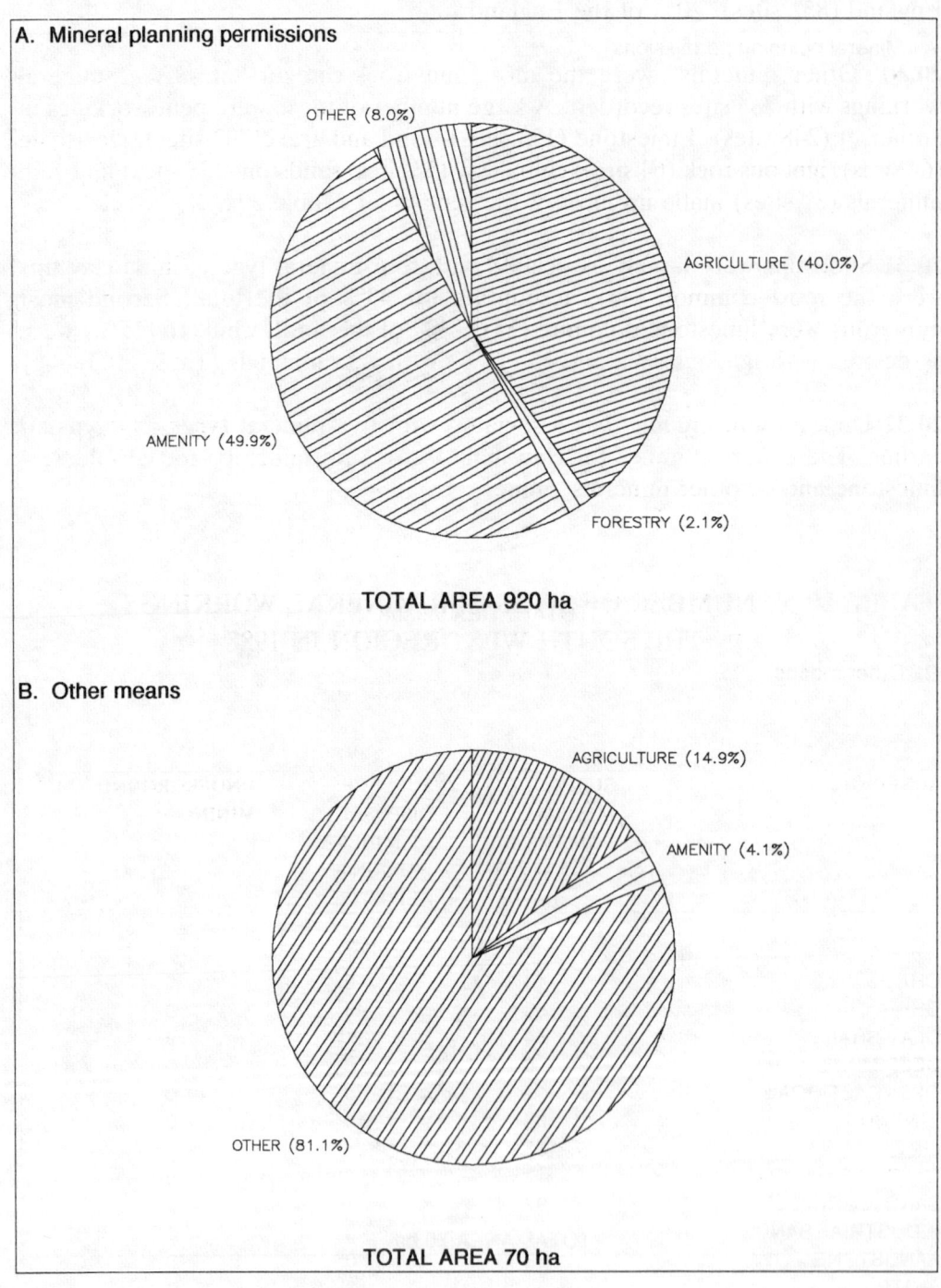

Number of Sites

20.28 There were 986 sites associated with mineral working in the region. Over 86% (852 sites) were surface mineral workings, 11% (106 sites) were for the surface disposal of mineral working deposits and 3% (28) were underground mines.

20.29 The region had the second largest number of surface mineral workings in England (852 sites), 20% of the England total.

20.30 "Other minerals" were the most numerous category of surface mineral workings with 284 sites recorded. A large number of these were peat workings in Somerset (248 sites). Limestone (153 sites), sand and gravel (92 sites), clay/shale (64 sites), igneous rock (64 sites) china clay (53 sites) sandstone (38 sites) and vein minerals (37 sites) made up most of the remainder (Table 20.5).

20.31 Spoil tips were largely associated with four mineral types. China clay tips were the most common (46), accounting for 43% of the total. Second most numerous were limestone/dolomite (25), 24% of the total, while 16 (15%) were associated with igneous rock and 13 (12%) with vein minerals (Table 20.5).

20.32 Underground mining was associated with five mineral types as discussed earlier. There were 9 mines for clay/shale, 8 for vein minerals, 6 coal mines, 4 limestone and 1 "other minerals" mine.

TABLE 20.5 NUMBER OF SITES FOR MINERAL WORKINGS IN THE SOUTH WEST REGION IN 1988.

MINERAL	SURFACE MINERAL WORKINGS	SURFACE DISPOSAL OF MINERAL WORKING DEPOSITS	UNDERGROUND MINING
CHALK	16	0	0
CHINA CLAY	53	46	0
CLAY/SHALE	64	1	9
DEEP MINED COAL	0	0	6
OPENCAST COAL	1	0	0
IGNEOUS ROCK	64	16	0
LIMESTONE	153	25	1
OIL/GAS EXPLORATION	17	0	0
SAND & GRAVEL	92	1	0
INDUSTRIAL SAND	8	0	0
SANDSTONE	38	0	0
SLATE	25	3	0
VEIN MINERALS	37	13	8
OTHER MINERALS	284	1	1
TOTAL	852	106	28

Land included in the Derelict Land Survey

20.33 There were 110 ha of land derelict as a result of mineral working in the region, less than 3% of mineral dereliction in England. This was equivalent to less than 1% of the total area in the region permitted for both surface mineral workings and spoil tips.

20.34 Surface mineral workings were the cause of 70% of the dereliction (80 ha). There were 30 ha of derelict spoil tips.

20.35 Over half of the dereliction was associated with former limestone quarries (60 ha). A further 22% were former sand and gravel sites (30 ha) (Figure 20.10).

FIGURE 20.10 AREA OF LAND DAMAGED BY MINERAL WORKINGS IN THE SOUTH WEST REGION AND INCLUDED IN THE 1988 DERELICT LAND SURVEY.

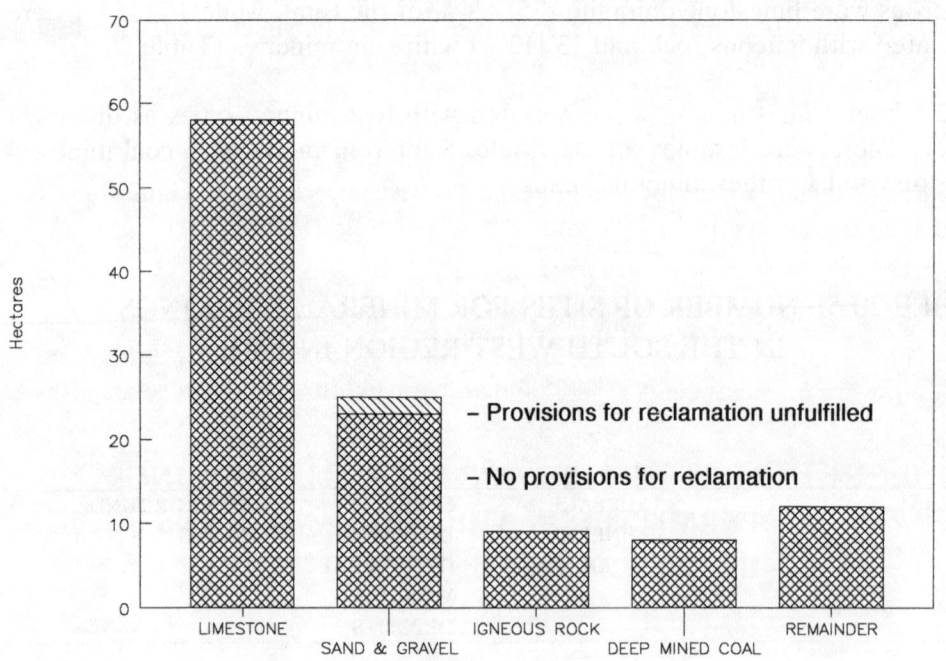

Reviews under Section 264A of the Town and Country Planning Act
20.36 Cornwall and Devon were the only authorities to have started their mineral reviews.

20.36 Seventy eight sites had been reviewed but at the time of the survey no further decisions had been taken.

21. WEST MIDLANDS

Introduction

21.1 The West Midlands consists of 11 mineral planning authorities; 4 shire counties (Hereford and Worcester, Shropshire, Staffordshire and Warwickshire) and 7 metropolitan borough councils (Birmingham City, Coventry City, Dudley, Sandwell, Solihull, Walsall and Wolverhampton). Minerals are worked in all authorities except Birmingham City.

21.2 There are five AONB's wholly or partly in the region; the Cotswold AONB, Malvern Hills AONB and the Wye Valley AONB in Hereford and Worcester; Shropshire Hills AONB in Shropshire; and Cannock Chase AONB in Staffordshire. Minerals were worked in all the AONB's except the Wye Valley. A small area of reclamation was recorded in the Wye Valley AONB.

Surface Mineral Workings

21.3 The West Midlands had the third smallest area of permissions for surface mineral workings in England. The total permitted area was 6490 ha, accounting for 7% of the England total. Despite this relatively small area, the region recorded a wide range of mineral types being worked; 10 mineral categories (including "other minerals").

21.4 Sand and gravel permissions were the most extensive accounting for 45% of the total area (2900 ha). Opencast coal (1060 ha), clay/shale (970 ha) and limestone (680 ha) made up a further 16%, 15% and 10% respectively (Figure 21.1).

21.5 Igneous rock permissions covered 2240 ha (7% of the region's total) and accounted for almost 20% of all permissions for this mineral in England (Table 21.1). Other minerals with significant contributions to the total permissions in England, were opencast coal (13%), sand and gravel (10%) and clay/shale (10%).

FIGURE 21.1   AREA OF SURFACE MINERAL WORKINGS PERMITTED IN THE WEST MIDLANDS REGION IN 1988 – MAIN MINERAL TYPES

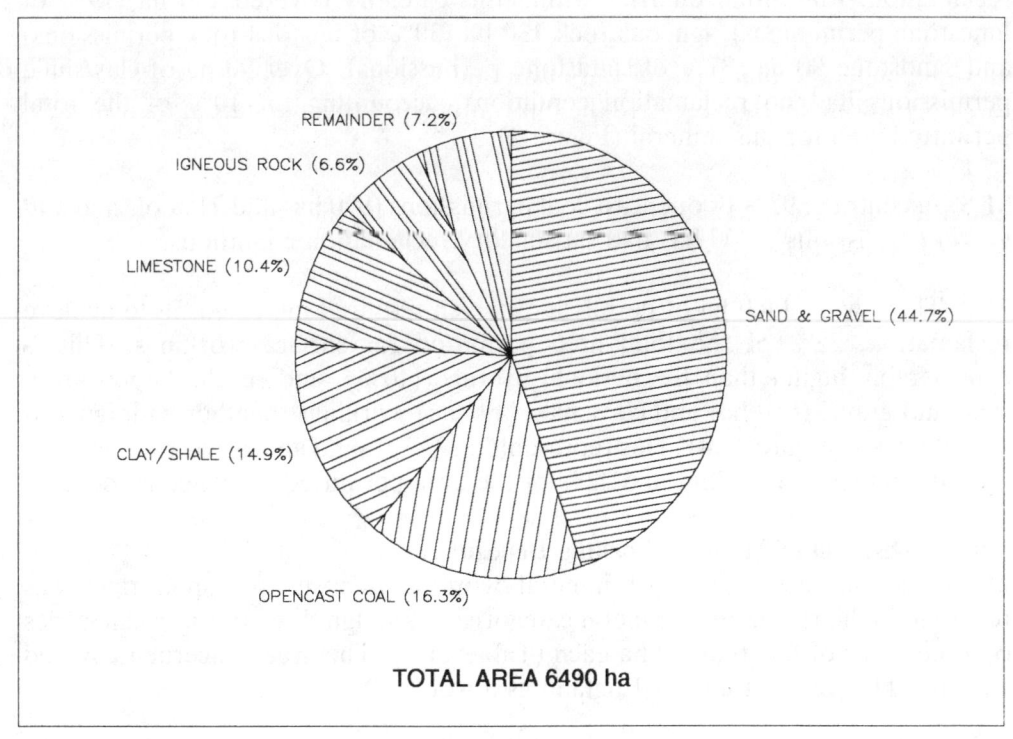

TABLE 21.1 TOTAL AREA PERMITTED FOR SURFACE MINERAL WORKINGS IN THE WEST MIDLANDS REGION, INDICATING THE PERCENTAGE REGIONAL CONTRIBUTION TO THE ENGLAND TOTAL.

MINERAL	TOTAL AREA PERMITTED IN REGION (ha)	TOTAL AREA PERMITTED IN ENGLAND (ha)	REGIONAL CONTRIBUTION TO ENGLAND TOTAL (%)
CLAY/SHALE	970	10090	9.6
OPENCAST COAL	1060	8420	12.6
GYPSUM/ANHYDRITE	10	810	1.2
IGNEOUS ROCK	430	2240	19.2
IRONSTONE	40	14420	0.3
LIMESTONE	680	11490	5.9
SAND & GRAVEL	2900	29040	10.0
INDUSTRIAL SAND	120	2130	5.6
SANDSTONE	248	2940	8.4
OTHER MINERALS	50	6530	0.8
TOTAL	6490		

Note: Areas are rounded to the nearest 10 hectares.

21.6 Two thirds of the permitted area (66% or 4250 ha) was considered to have satisfactory reclamation conditions (Figure 21.2). A further 1490 ha (23%) had reclamation conditions which were considered to be unsatisfactory. Less than 9% (550 ha) had no provisions for reclamation whilst 3% (190 ha) had been damaged and was unlikely to be reclaimed because provisions for reclamation were unfulfilled.

21.7 Hard rock quarries in particular had the largest areas with no provisions for reclamation. Limestone quarries within this category covered 150 ha (19% of limestone permissions), igneous rock 130 ha (30% of igneous rock permissions) and sandstone 90 ha (37% of sandstone permissions). Over 90 ha of clay/shale permissions had no reclamation conditions, accounting for 10% of the total permitted area for this mineral (Figure 21.3).

21.8 In contrast, 93% of opencast coal permissions (990 ha) and 71% of sand and gravel permissions (2060 ha) had satisfactory reclamation conditions.

21.9 There were 1460 ha of permissions which required imported fill to achieve reclamation, 22% of the total area permitted for surface workings. This is considerably higher than the national average (10%), and results largely from sand and gravel (650 ha) and clay/shale (490 ha) workings together with igneous rock quarries (Figure 21.4). This represented 22% of sand and gravel permissions, 50% of clay/shale workings and 42% of the area permitted for igneous rock.

Surface Disposal of Mineral Working Deposits
21.10 The surface disposal of mineral working deposits (or spoil tips) was recorded in the region for 8 mineral categories, although three of these categories covered areas of less than 10 ha each (Table 21.2). The area concerned covered 1050 ha, 11% of the total for England as a whole.

FIGURE 21.2 NATURE OF PROVISIONS FOR RECLAMATION FOR SURFACE MINERAL WORKINGS IN THE WEST MIDLANDS REGION IN 1988.

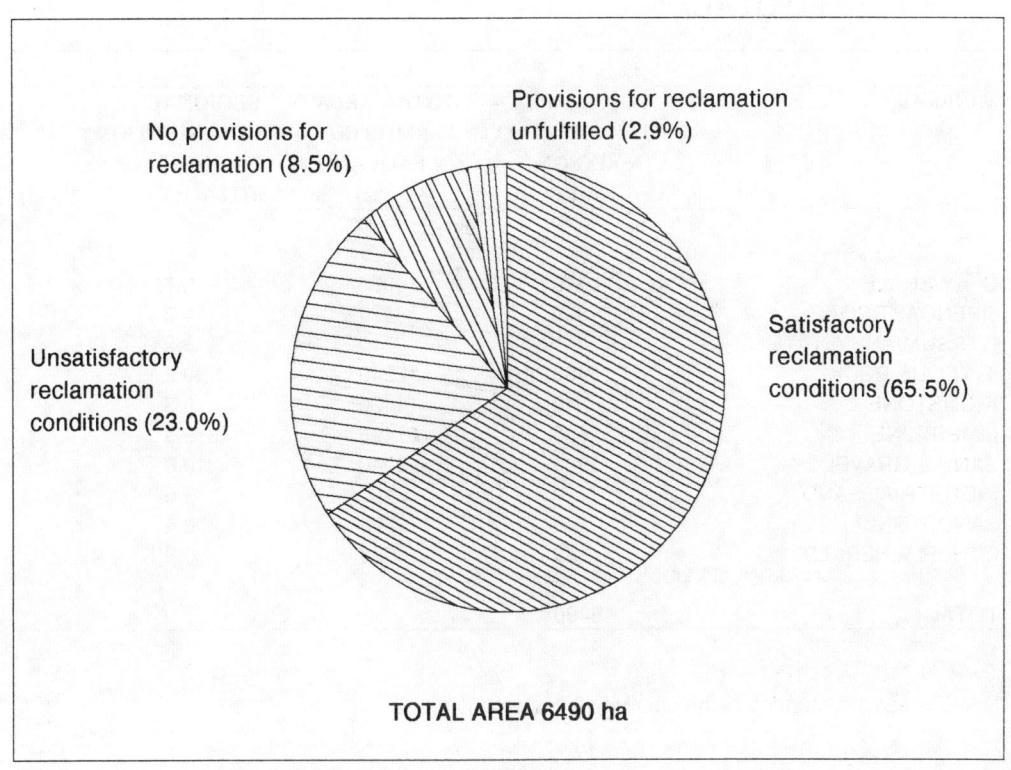

FIGURE 21.3 AREA AND NATURE OF PROVISIONS FOR RECLAMATION OF SURFACE MINERAL WORKINGS IN THE WEST MIDLANDS REGION IN 1988 – MAIN MINERAL TYPES.

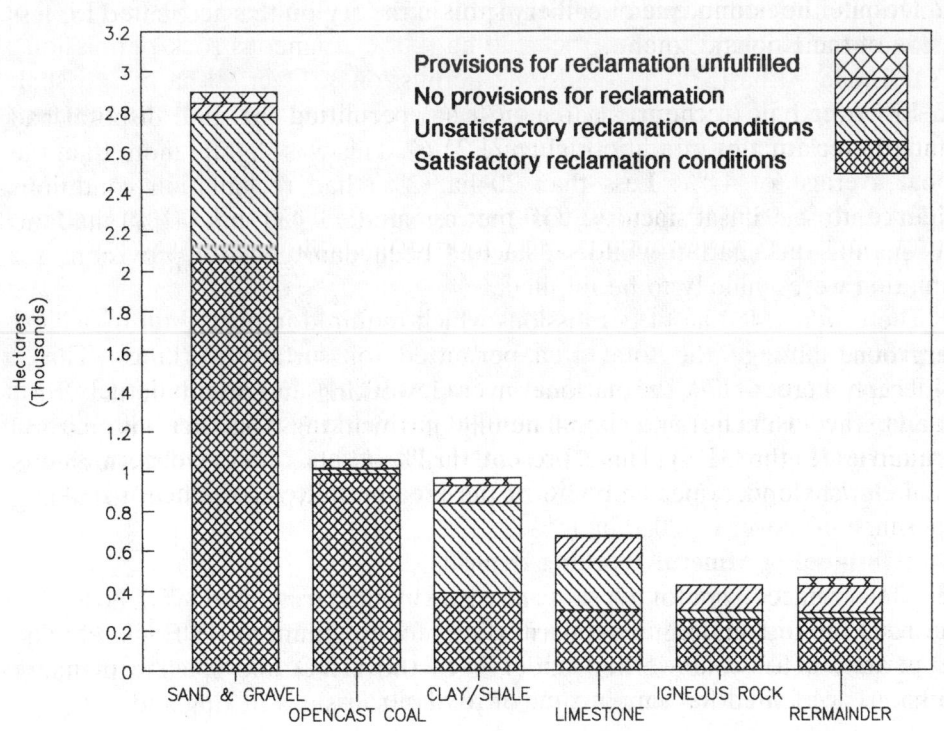

141

FIGURE 21.4 SURFACE MINERAL WORKINGS DEPENDENT ON IMPORTED FILL IN THE WEST MIDLANDS REGION IN 1988 – MAIN MINERAL TYPES

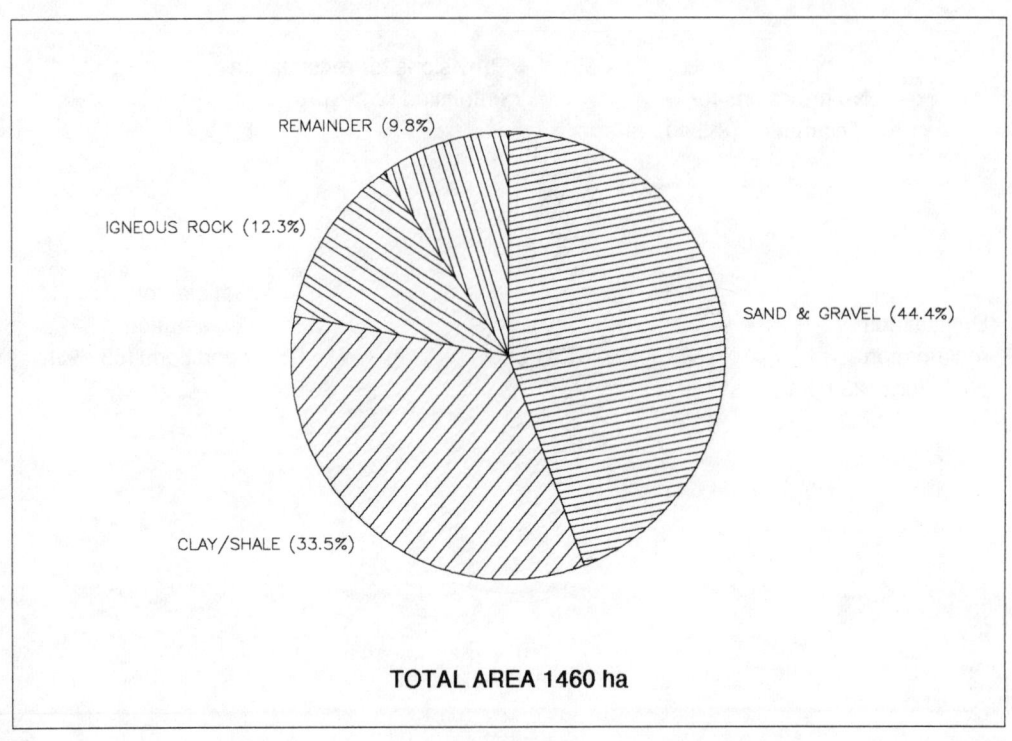

21.11 Spoil tips from deep mined coal accounted for 83% of the area (870 ha) as shown in Figure 21.5.

21.12 Despite the dominance of colliery spoil in the region this accounted for less than 8% of the England total.

21.13 Just over half of the total area (530 ha) permitted for spoil disposal had satisfactory reclamation conditions (Figure 21.6). This was slightly more than the national average of 47%. Less than 20 ha (2%) had reclamation conditions considered to be unsatisfactory. Of the remainder, 430 ha (41%) had no provisions for reclamation whilst 80 ha had been damaged but provisions for reclamation were unlikely to be fulfilled.

Underground mining
21.14 There were 41170 ha of underground workings recorded in the West Midlands. The vast majority was coal mining of which 31380 ha (76%) is or could be worked under the GDO (Table 21.3). A further 4450 ha (11%) of the area was for coal worked under specific planning permissions. Gypsum/anhydrite mining had permissions covering 2030 ha (5%).

21.15 The permitted area for gypsum/anhydrite mining accounted for almost 14% of the total area of underground permissions for this mineral in England. The areas permitted for coal worked under both the GDO and specific planning permissions accounted for almost 10% of total permissions in England.

FIGURE 21.5 AREA PERMITTED FOR THE SURFACE DISPOSAL OF MINERAL WORKING DEPOSITS IN THE WEST MIDLANDS REGION IN 1988 – MAIN MINERAL TYPES.

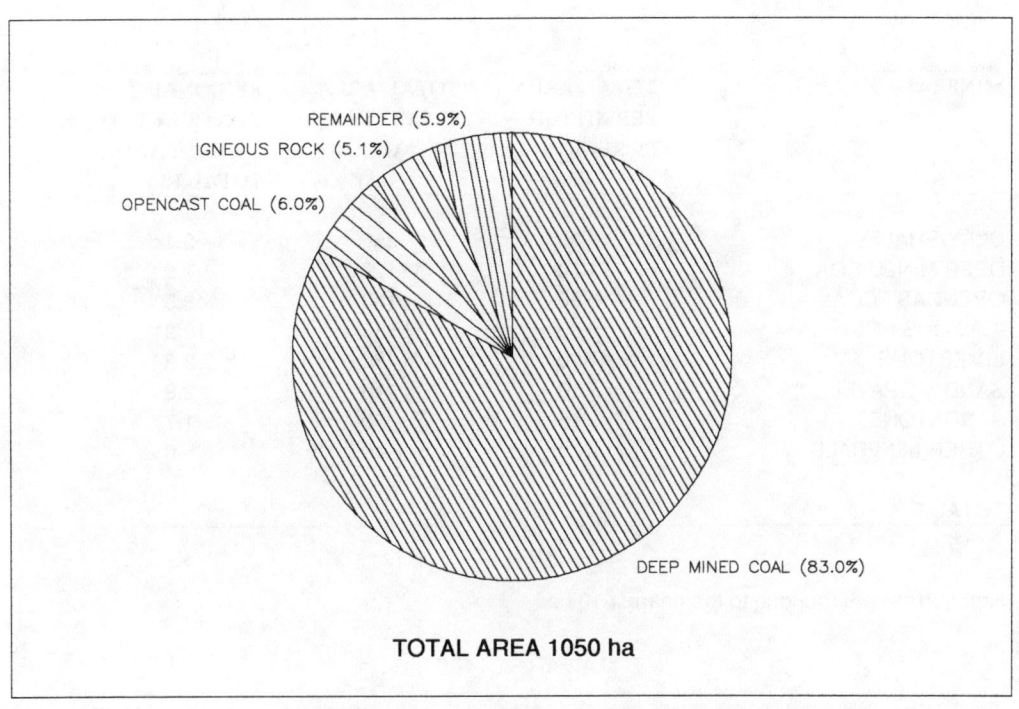

FIGURE 21.6 NATURE OF PROVISIONS FOR RECLAMATION FOR THE SURFACE DISPOSAL OF MINERAL WORKING DEPOSITS IN THE WEST MIDLANDS REGION IN 1988.

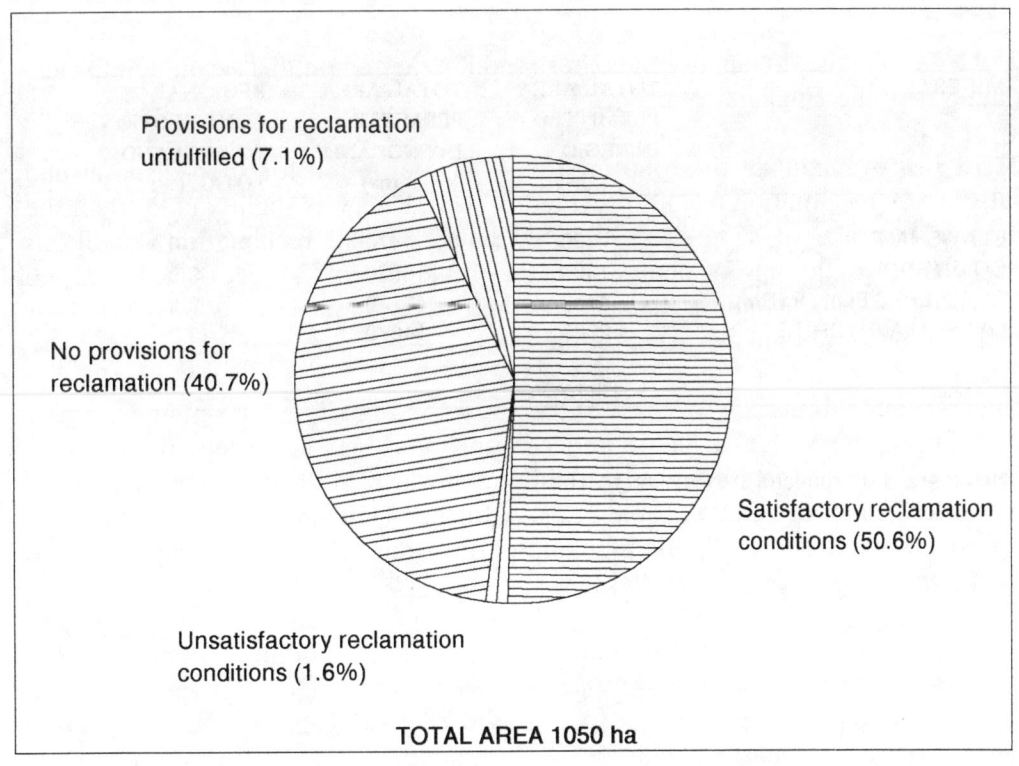

TABLE 21.2 TOTAL AREA PERMITTED FOR THE SURFACE DISPOSAL OF MINERAL WORKING DEPOSITS IN THE WEST MIDLANDS REGION, INDICATING THE PERCENTAGE REGIONAL CONTRIBUTION TO THE ENGLAND TOTAL.

MINERAL	TOTAL AREA PERMITTED IN REGION (ha)	TOTAL AREA PERMITTED IN ENGLAND (ha)	REGIONAL CONTRIBUTION TO ENGLAND TOTAL (%)
CLAY/SHALE	30	350	9.1
DEEP MINED COAL	870	11120	7.9
OPENCAST COAL	60	130	48.5
IGNEOUS ROCK	50	530	10.2
LIMESTONE	20	780	2.9
SAND & GRAVEL	<10	140	2.8
SANDSTONE	<10	60	1.7
OTHER MINERALS	<10	40	5.0
TOTAL	1052		

Note: Areas are rounded to the nearest 10 ha.

TABLE 21.3 TOTAL AREA OF UNDERGROUND PERMISSIONS IN THE WEST MIDLANDS REGION, INDICATING THE PERCENTAGE REGIONAL CONTRIBUTION TO THE ENGLAND TOTAL.

MINERAL	TOTAL AREA PERMITTED IN REGION (ha)	TOTAL AREA PERMITTED IN ENGLAND (ha)	REGIONAL CONTRIBUTION TO ENGLAND TOTAL (%)
CLAY/SHALE	20	1470	1.4
COAL (GDO)	31380	315820	9.9
COAL (SPEC PLAN PERM)	4450	46380	9.6
GYPSUM/ANHYDRITE	2030	14670	13.8
TOTAL	41170		

Note: Areas are rounded to the nearest 10 ha.

Land covered by permissions which include aftercare conditions

21.16 There were 112 permissions which included aftercare conditions covering a total area of 3370 ha. Surface mineral workings accounted for 101 of the permissions which covered 3060 ha. Spoil disposal areas with aftercare conditions covered 310 ha involving 11 permissions.

21.17 The majority of the area covered by aftercare conditions was associated with sand and gravel and opencast coal workings (Figure 21.7). Sand and gravel permissions accounted for 40% of the total area (1340 ha) whilst opencast coal permissions covered 35% (1170 ha). Permissions for clay/shale workings covered 270 ha (8%) and limestone quarries 110 ha (3%).

21.18 Agriculture was the intended afteruse for 71% of the area with aftercare conditions (2380 ha). Amenity afteruses were proposed for 16% (550 ha) and forestry 13% (450 ha) (Figure 21.7).

21.19 Most of the areas proposed for forestry were sand and gravel workings (370 ha) which accounted for 84% of the land with this proposed afteruse. A further 9% (40 ha) of forestry was proposed for spoil disposal areas from deep mined coal. All mineral types worked in the region (except gypsum/anhydrite which only had 1 ha with aftercare conditions) had some area with amenity afteruses proposed. The largest areas were on opencast coal (170 ha) and sand and gravel (160 ha) workings, largely reflecting the large areas of these mineral types with aftercare conditions.

Reworking of Mineral working deposits

21.20 Three mineral planning authorities had mineral working deposits being reworked in their area (Shropshire, Staffordshire and Warwickshire). Two mineral categories were involved; deep mined coal and "other minerals" (see Table J2 in Volume 2).

21.21 Only four sites were involved of which three were worked under specific planning permissions. Three colliery spoil tips had permissions for reworking, covering an area of 22 ha.

Reclamation

21.22 Land reclaimed after mineral workings between 1982 and 1988 in the region covered 2080 ha, 10% of the total area of mineral workings reclaimed in England.

21.23 Over 86% of this area (1800 ha) was reclaimed by mineral planning permissions and 280 ha (14%) were reclaimed by "other means".

21.24 Former sand and gravel and opencast coal sites accounted for most of the land reclaimed under mineral planning permissions. Sand and gravel reclamation accounted for 53% of the total (960 ha), opencast coal 28% (500 ha). The reclamation of clay/shale sites added a further 120 ha (7%) (Figure 21.8).

21.25 Over three quarters of the land reclaimed by "other means" were former clay/shale sites (220 ha) (Figure 21.8). Most of the remainder, were former sand and gravel sites (20% or 60 ha).

21.26 Agriculture was by far the most common afteruse of land reclaimed by mineral planning permissions (Figure 21.9), accounting for 71% of the total (1270 ha). Amenity afteruses were the second most common (17%) covering an area of 320 ha. A further 110 ha were reclaimed to "other uses" (6%) and 90 ha (5%) were reclaimed to forestry.

21.27 Very little of the area reclaimed by "other means" was reclaimed to agriculture, (less than 30 ha - 10% of land reclaimed by this route) (Figure 21.9).

FIGURE 21.7 AREA OF MINERAL WORKINGS COVERED BY AFTERCARE CONDITIONS IN THE WEST MIDLANDS REGION IN 1988 – MAIN MINERAL TYPES, AND PROPOSED ENDUSES.

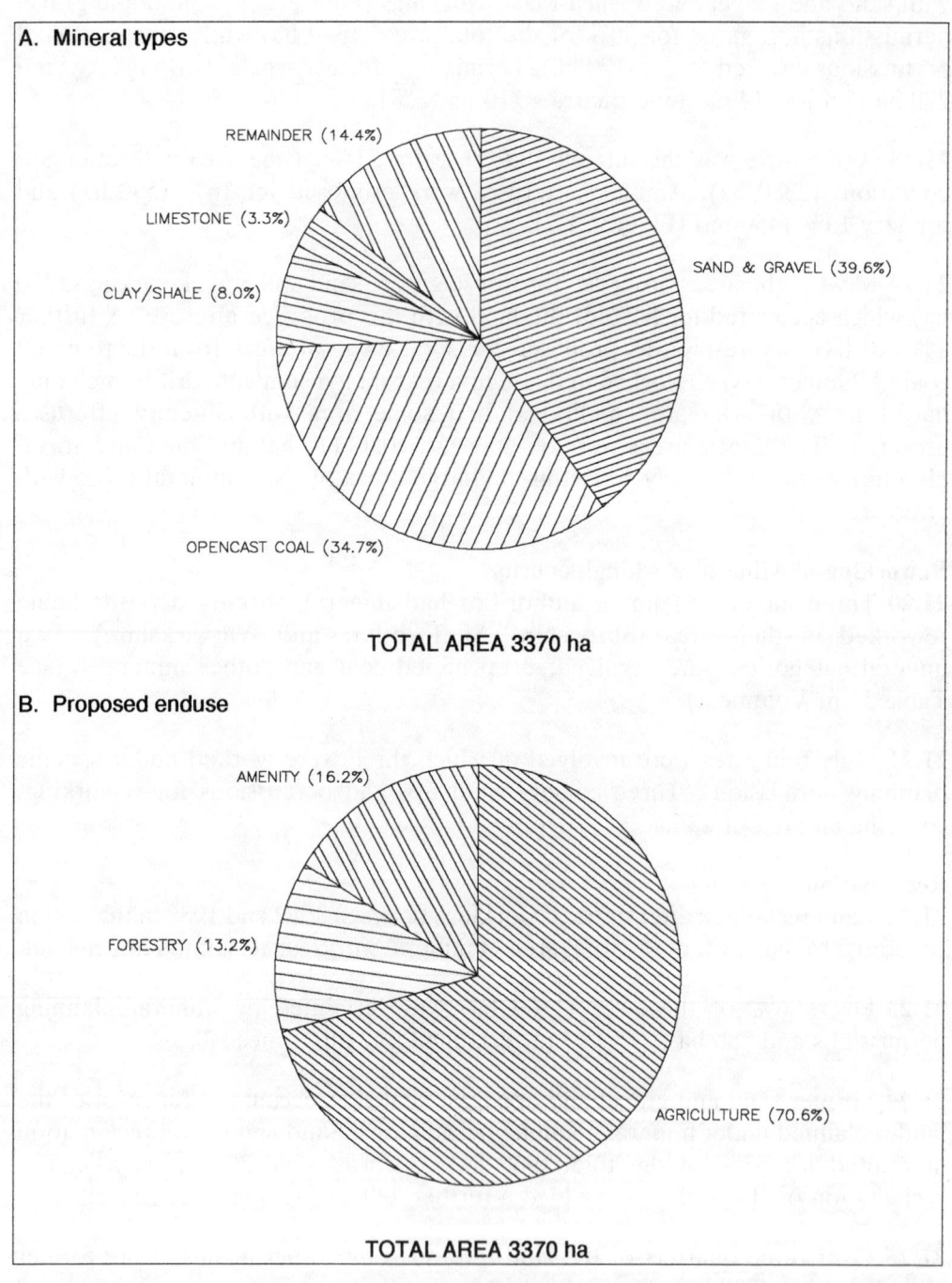

FIGURE 21.8 AREA OF LAND RECLAIMED BY MINERAL PLANNING PERMISSIONS (A) AND BY OTHER MEANS (B) IN THE WEST MIDLANDS REGION BETWEEN 1982 AND 1988 – MAIN MINERAL TYPES.

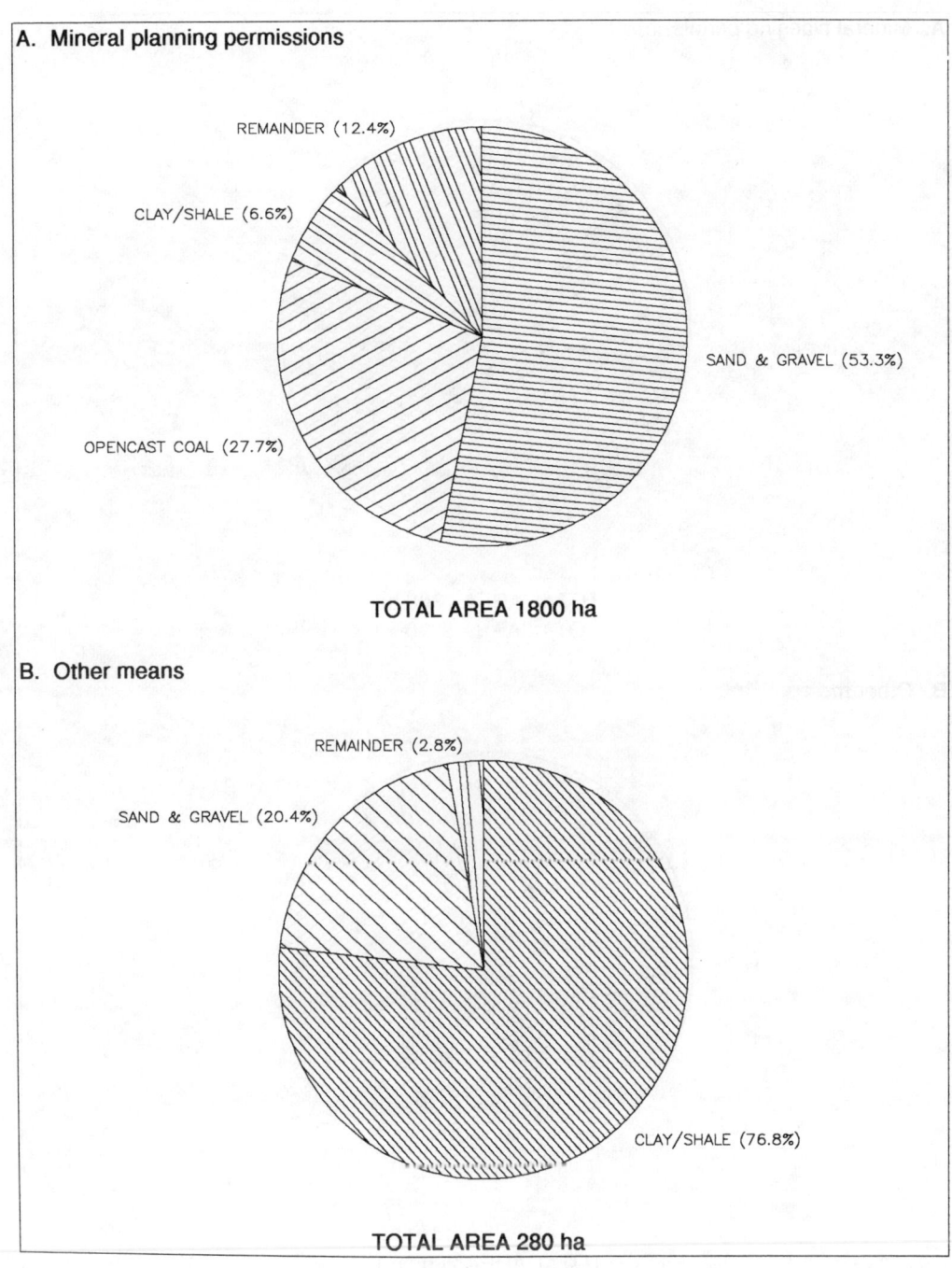

FIGURE 21.9 CHOSEN AFTERUSE OF LAND RECLAIMED BY MINERAL PLANNING PERMISSIONS (A) AND BY OTHER MEANS (B) IN THE WEST MIDLANDS REGION BETWEEN 1982 AND 1988.

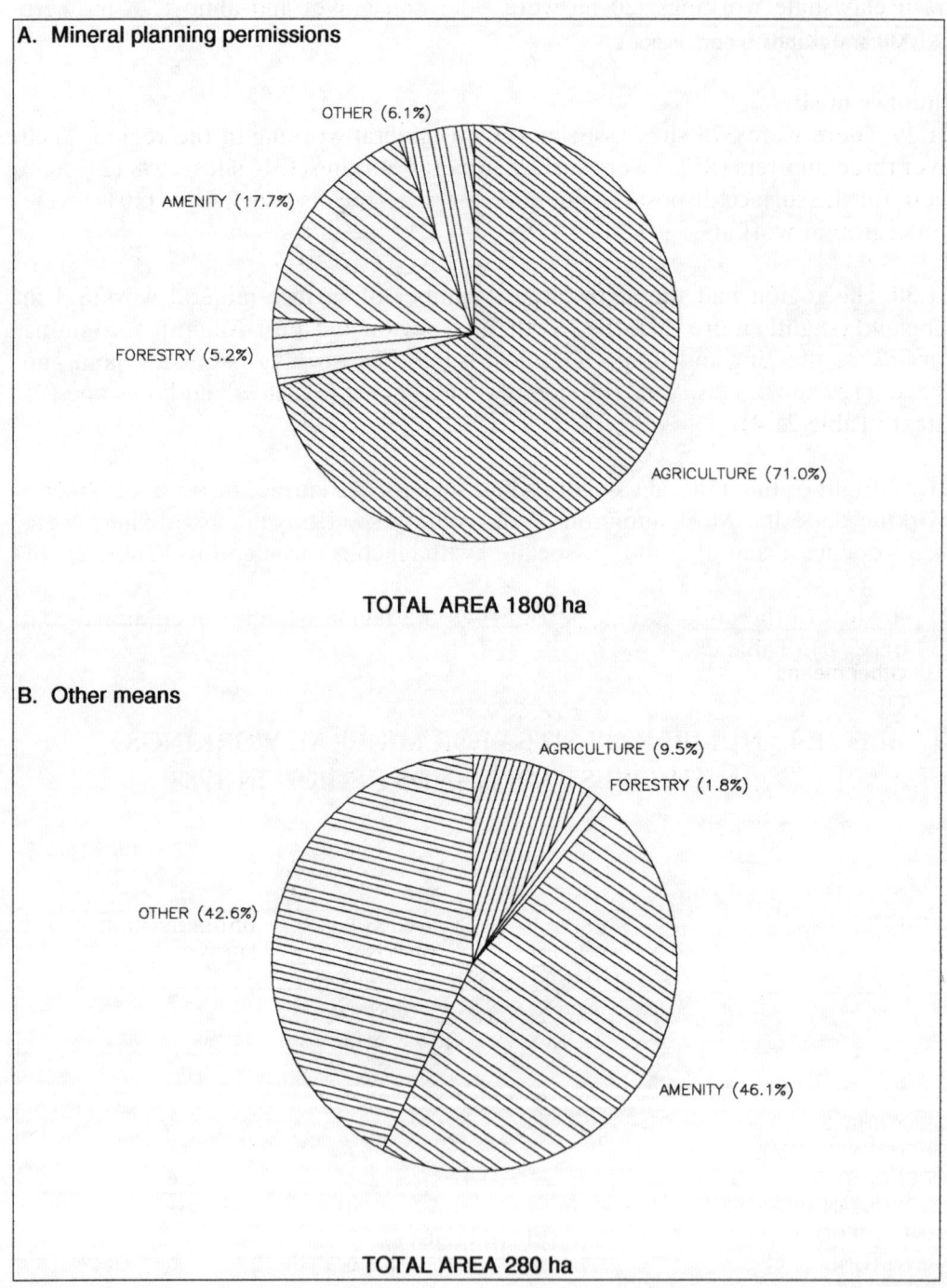

Amenity afteruses were the most common accounting for 46% of the total (130 ha), a slightly larger area than was reclaimed to "other uses" (120 ha or 43% of the total). Less than 10 ha were reclaimed to forestry (2%).

21.28 Almost 96% of land reclaimed by both mineral planning permissions and by "other means" was considered to be satisfactory - slightly above the national average (95%). Of the 80 ha reclaimed to an unsatisfactory standard, 30 ha were from clay/shale workings, 20 ha were sand and gravel and almost 20 ha were igneous rock sites.

Number of sites

21.29 There were 373 sites associated with mineral working in the region. Well over three quarters (85%) were surface mineral workings (316 sites), 6% (21 sites) were for the surface disposal of mineral working deposits and 36 sites (10%) were underground workings.

21.30 The region had the sixth largest number of surface mineral workings in England (slightly more than the North West region and East Anglia), accounting for 7% of the England total. The most numerous workings were for sand and gravel (129 sites), clay/shale (80 sites), opencast coal (35 sites) and limestone (25 sites) (Table 21.4).

21.31 Eight of the mineral categories had sites for the surface disposal of mineral working deposits. Most numerous were colliery spoil tips (7 sites). There were also 5 opencast coal sites and 3 associated with igneous rock deposits (Table 21.4).

21.32 Most of the underground workings in the region were for deep mined coal (34 or 85%) (Table 21.4).

TABLE 21.4 NUMBER OF SITES FOR MINERAL WORKINGS IN THE WEST MIDLANDS REGION IN 1988.

MINERAL	SURFACE MINERAL WORKINGS	SURFACE DISPOSAL OF MINERAL WORKING DEPOSITS	UNDERGROUND MINING
CLAY/SHALE	80	1	1
DEEP MINED COAL	0	7	34
OPENCAST COAL	35	5	0
GYPSUM/ANHYDRITE	1	0	1
IGNEOUS ROCK	15	3	0
IRONSTONE	2	0	0
LIMESTONE	25	2	0
SAND & GRAVEL	129	1	0
INDUSTRIAL SAND	6	0	0
SANDSTONE	18	1	0
OTHER MINERALS	5	1	0
TOTAL	316	21	36

Land included in the Derelict Land Survey
21.33 There were 290 ha of land recorded in both this and the 1988 Derelict Land Survey. This was less than 7% of the total area recorded as derelict in England, and 4% of the total permitted area of surface workings and spoil tips in the region.

21.34 Surface mineral workings accounted for 54% of the dereliction (150 ha), while mineral working deposits accounted for 46% (140 ha). Over 65% of dereliction (190 ha) resulted from provisions for reclamation being unfulfilled. The remaining 100 ha (35%) were derelict because of a lack of provisions for reclamation.

21.35 The mineral type with the largest derelict area was colliery spoil tips which accounted for 38% of the total (Figure 21.10). Slightly over half of this dereliction was a result of provisions for reclamation being unfulfilled. Sand and gravel, clay/shale and opencast coal also contributed to the total (60 ha, 50 ha and 40 ha respectively).

FIGURE 21.10 AREA OF LAND DAMAGED BY MINERAL WORKINGS IN THE WEST MIDLANDS AND INCLUDED IN THE 1988 DERELICT LAND SURVEY

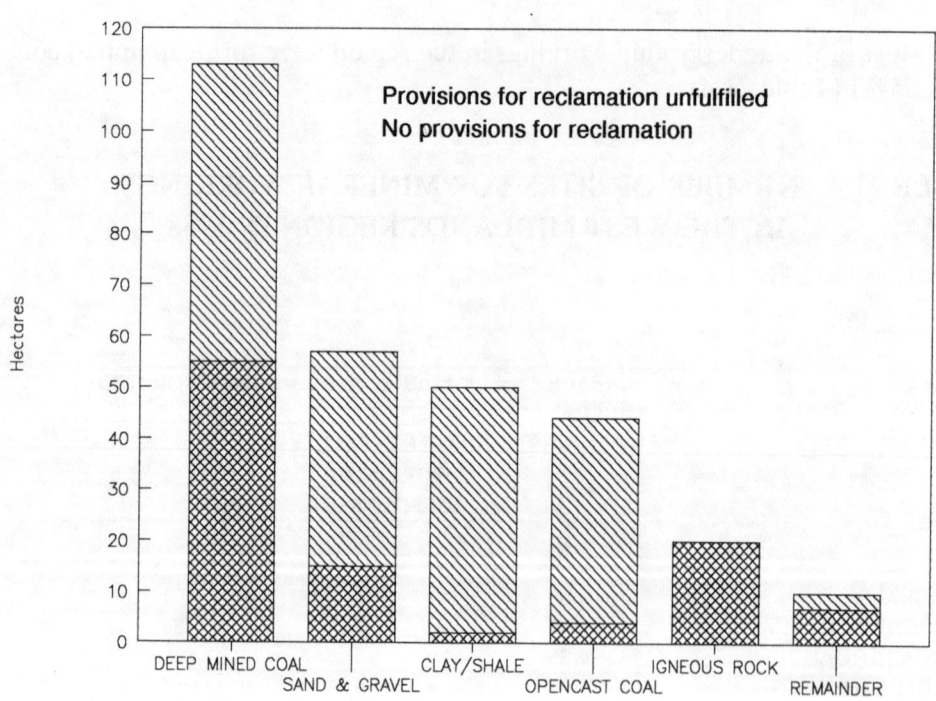

Reviews under Section 264A of the Town and Country Planning Act 1971
21.36 No mineral planning authorities within the region had started their review procedures by 1st April 1988.

22. NORTH WEST REGION

Introduction

22.1 The North West region contains 16 mineral planning authorities; 2 shire counties (Cheshire and Lancashire) and 14 metropolitan borough councils within Greater Manchester (Bolton, Bury, Oldham, Rochdale, Salford City, Stockport, Tameside, Trafford and Wigan) and Merseyside (Knowsley, Liverpool City, St Helens, Sefton and Wirral). Mineral workings were recorded in all mineral planning authorities except Liverpool City.

22.2 There are two AONB's which are partly in the region, both in Lancashire; Arnside and Silverdale AONB and Forest of Bowland AONB. Mineral workings were recorded in both AONB's.

Surface Mineral Workings

22.3 The North West region recorded the smallest area permitted for surface mineral workings in England, 4410 ha, less than 5% of the total area permitted in England.

22.4 Despite the relatively small area permitted, eight mineral categories were recorded (including "other minerals") (Table 22.1).

TABLE 22.1 TOTAL AREA PERMITTED FOR SURFACE MINERAL WORKINGS IN THE NORTH WEST REGION, INDICATING THE PERCENTAGE REGIONAL CONTRIBUTION TO THE ENGLAND TOTAL.

MINERAL	TOTAL AREA PERMITTED IN REGION (ha)	TOTAL AREA PERMITTED IN ENGLAND (ha)	REGIONAL CONTRIBUTION TO ENGLAND TOTAL (%)
CLAY/SHALE	660	10090	6.5
OPENCAST COAL	160	8420	1.9
LIMESTONE	240	11490	2.1
OIL/GAS EXPLORATION	<10	80	5.0
SAND & GRAVEL	830	29040	2.9
INDUSTRIAL SAND	700	2130	32.9
SANDSTONE	1190	2940	40.5
OTHER MINERALS	630	6530	9.6
TOTAL	4410		

Note: Areas are rounded to the nearest 10 hectares.

22.5 Sandstone permissions were the most extensive accounting for 27% of the total (1190 ha). Sand and gravel permissions covered 830 ha (19%), industrial sand 700 ha (15%) and clay/shale 660 ha (15%) (Figure 22.1).

22.6 The area permitted for sandstone quarrying represented over 40% of the total area permitted for the extraction of this mineral in England (Table 22.1). Permissions for industrial sand extraction in the region accounted for 33% of the England total.

22.7 More than half (52%) of the area permitted for surface mineral workings was considered to have satisfactory reclamation conditions (2280 ha) (Figure 22.2). A further 17% (750 ha) had reclamation conditions which were considered to be unsatisfactory. Land with no provisions for reclamation accounted for 25% of the permitted area (1120 ha) and land damaged but unlikely to be reclaimed because provisions for reclamation were unfulfilled covered 260 ha (6%).

22.8 All opencast coal permissions (160 ha), 98% of industrial sand permissions (680 ha) and 83% of sand and gravel permissions (690 ha) had satisfactory reclamation conditions (Figure 22.3). In contrast, only 18% of the permitted area for sandstone quarrying had satisfactory reclamation conditions, while 42% had no provisions for reclamation (500 ha) and 26% had unsatisfactory reclamation conditions (310 ha). Clay/shale and "other minerals" sites also had only a small proportion covered by satisfactory reclamation conditions, 43% (290 ha) and 39% (250 ha) respectively.

22.9 There were only 290 ha which required imported fill to achieve reclamation in the region, less than 7% of the permitted area for surface mineral workings. Half of this area was associated with clay/shale sites (140 ha), 28% (80 ha) with sandstone quarries and the remaining 22% (60 ha) with sand and gravel workings (Figure 22.4).

FIGURE 22.1   AREA OF SURFACE MINERAL WORKINGS PERMITTED IN THE NORTH WEST REGION IN 1988 – MAIN MINERAL TYPES.

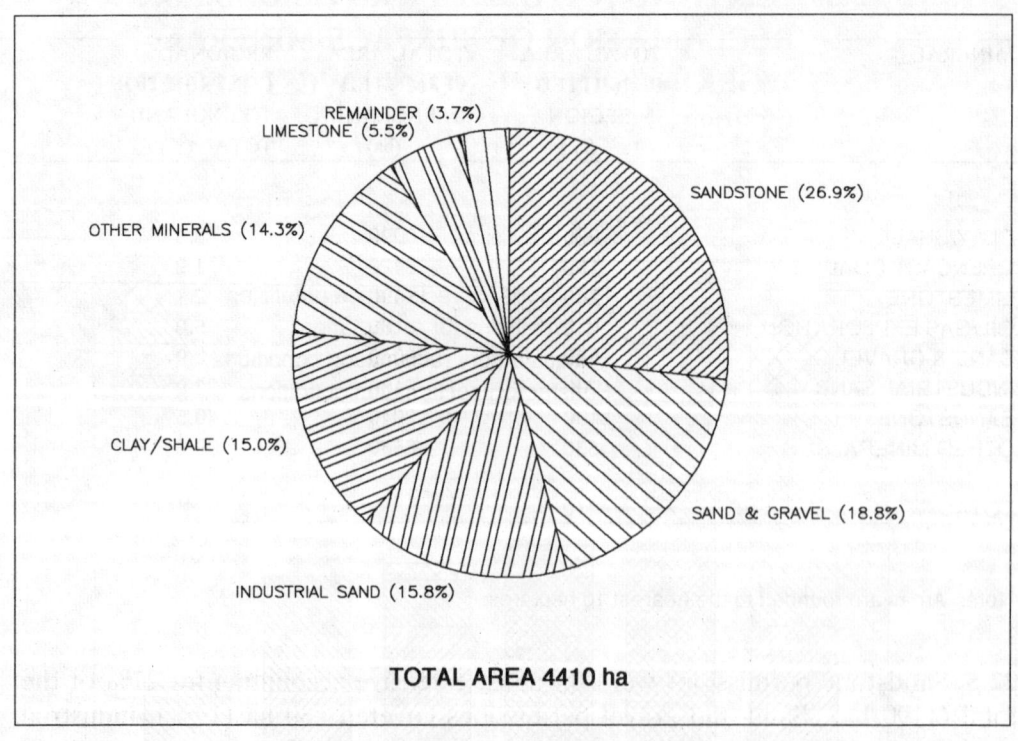

FIGURE 22.2 NATURE OF PROVISIONS FOR RECLAMATION FOR SURFACE MINERAL WORKINGS IN THE NORTH WEST REGION IN 1988.

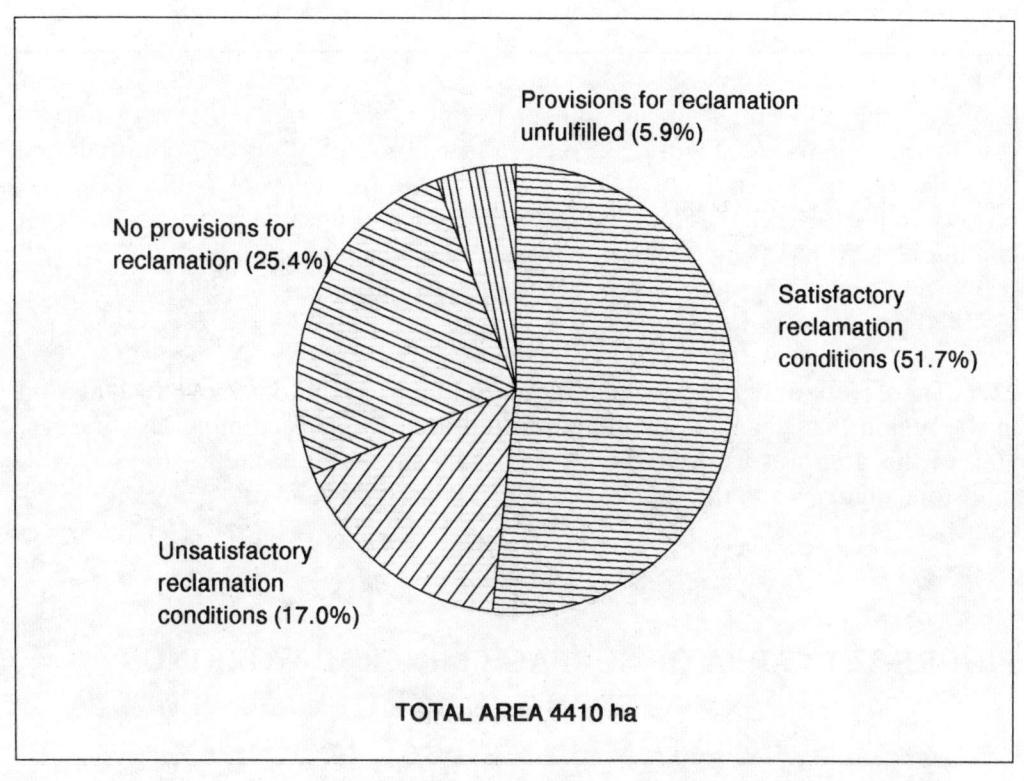

FIGURE 22.3 AREA AND NATURE OF PROVISIONS FOR RECLAMATION OF SURFACE MINERAL WORKINGS IN THE NORTH WEST REGION IN 1988 – MAIN MINERAL TYPES.

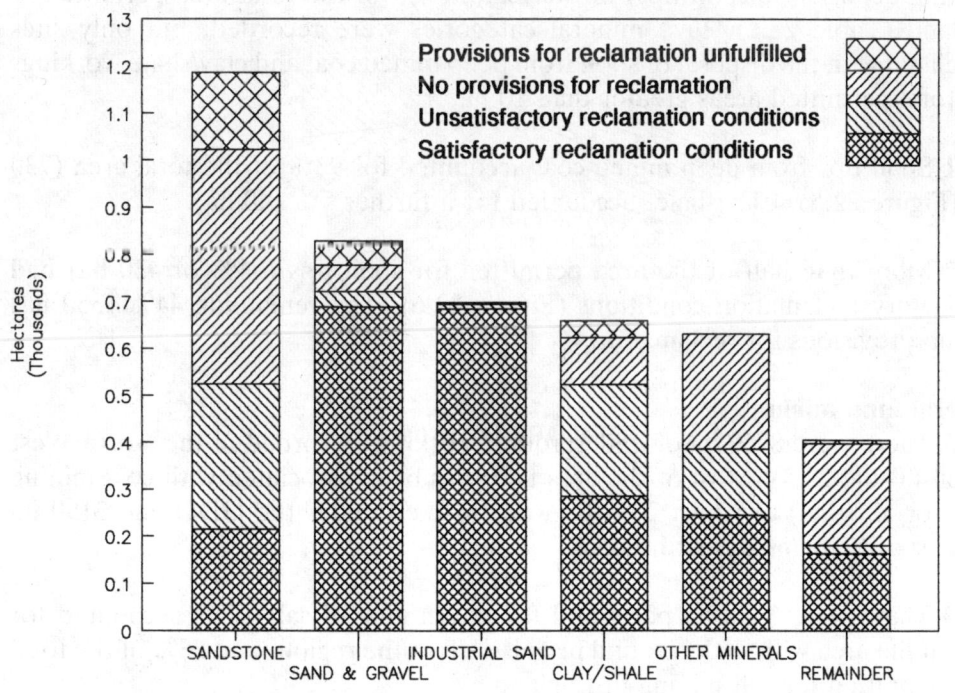

FIGURE 22.4 SURFACE MINERAL WORKINGS DEPENDENT ON IMPORTED FILL IN THE NORTH WEST REGION IN 1988.

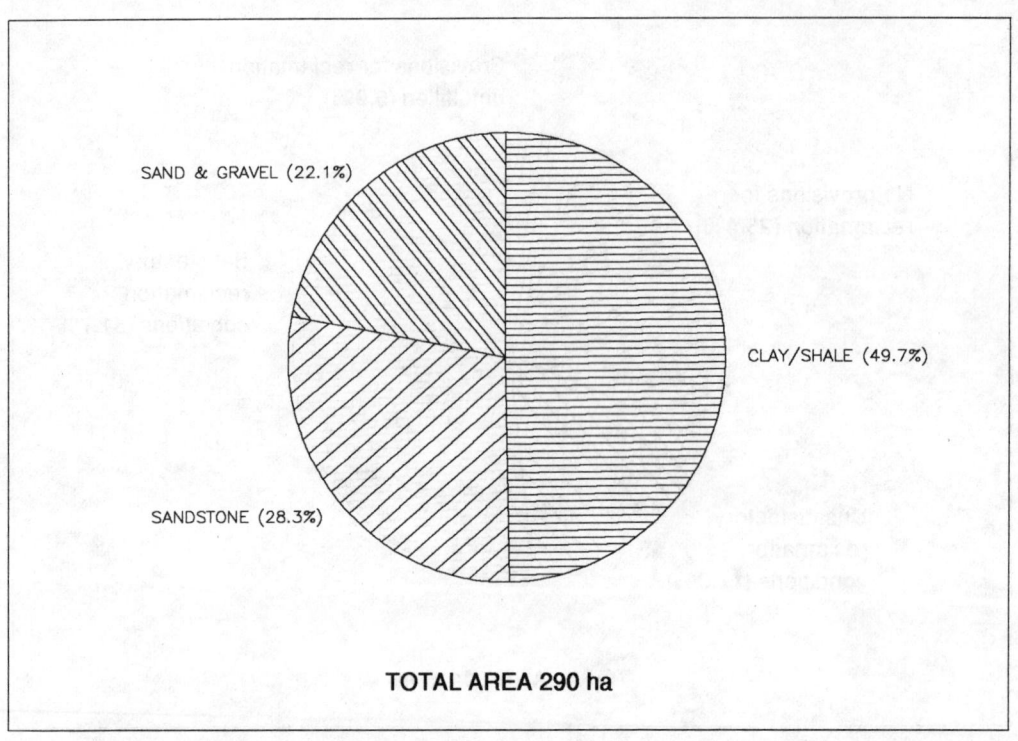

Surface Disposal of Mineral Working Deposits

22.10 There where 790 ha with permissions for the surface disposal of mineral working deposits (ie spoil tips) in the region, 4% of the total area permitted in England (Table 22.2). Five mineral categories were recorded, but only sites associated with the disposal of spoil from deep mined coal and clay/shale workings had total permitted areas greater than 10 ha.

22.11 Spoil tips from deep mined coal accounted for 93% of the total area (730 ha) (Figure 22.5). Clay/shale, accounted for a further 5% (40 ha).

22.12 More than half of the area permitted for spoil tips (54% or 430 ha) had satisfactory reclamation conditions (Figure 22.6). The remaining 44% (350 ha) had no provisions for reclamation.

Underground mining
22.13 There were 8750 ha of underground workings recorded in the North West region (Table 22.3). Most of the area is or has been associated with coal mining (60% or 5270 ha) of which 2080 ha were worked under the GDO and 3190 ha under specific planning permissions.

22.14 There were 3450 ha permitted for the mining of salt. This accounted for 39% of the area with underground permissions in the region, and 76% of the total area permitted for salt mining in England.

FIGURE 22.5 AREA PERMITTED FOR THE SURFACE DISPOSAL OF MINERAL WORKING DEPOSITS IN THE NORTH WEST REGION IN 1988 – MAIN MINERAL TYPES.

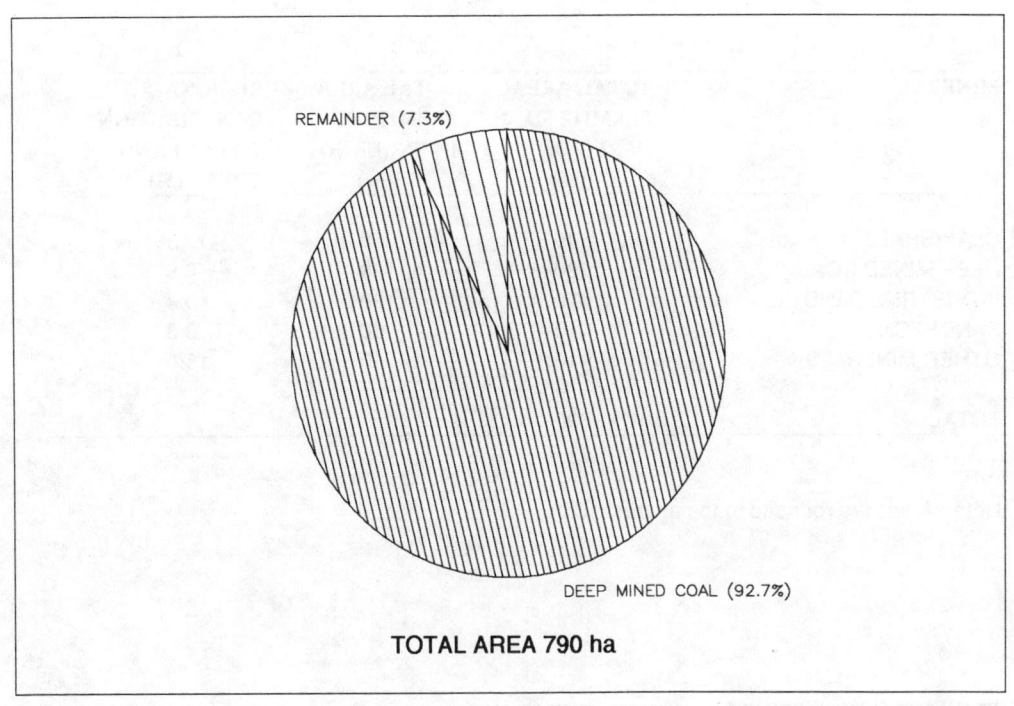

FIGURE 22.6 NATURE OF PROVISIONS FOR RECLAMATION FOR THE SURFACE DISPOSAL OF MINERAL WORKING DEPOSITS IN THE NORTH WEST REGION IN 1988.

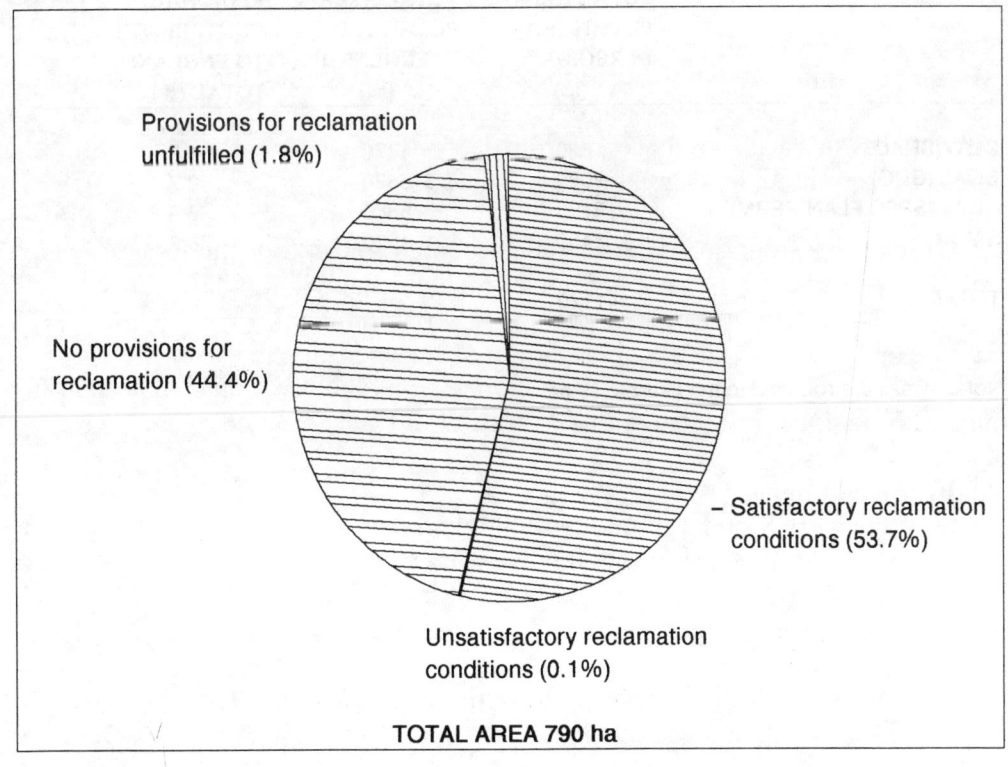

TABLE 22.2 TOTAL AREA PERMITTED FOR THE SURFACE DISPOSAL OF MINERAL WORKING DEPOSITS IN THE NORTH WEST REGION, INDICATING THE PERCENTAGE REGIONAL CONTRIBUTION TO THE ENGLAND TOTAL.

MINERAL	TOTAL AREA PERMITTED IN REGION (ha)	TOTAL AREA PERMITTED IN ENGLAND (ha)	REGIONAL CONTRIBUTION TO ENGLAND TOTAL (%)
CLAY/SHALE	40	350	12.3
DEEP MINED COAL	730	11120	6.6
INDUSTRIAL SAND	<10	130	5.4
SANDSTONE	<10	60	3.3
OTHER MINERALS	<10	40	15.0
TOTAL	790		

Note: Areas are rounded to the nearest 10 ha.

TABLE 22.3 TOTAL AREA OF UNDERGROUND PERMISSIONS IN THE NORTH WEST REGION, INDICATING THE PERCENTAGE REGIONAL CONTRIBUTION TO THE ENGLAND TOTAL.

MINERAL	TOTAL AREA PERMITTED IN REGION (ha)	TOTAL AREA PERMITTED IN ENGLAND (ha)	REGIONAL CONTRIBUTION TO ENGLAND TOTAL (%)
CLAY/SHALE	30	1470	1.8
COAL (GDO)	2080	315820	0.7
COAL (SPEC PLAN PERM)	3190	46380	6.9
SALT	3450	4520	76.2
TOTAL	41170		

Note: Areas are rounded to the nearest 10 ha.

Land covered by permissions which include aftercare conditions
22.15 There were 32 permissions which included aftercare conditions, covering a total area of 680 ha, equivalent to 15% of the region's current permissions. This was below the national average of 18%, and the smallest total area covered by aftercare conditions in any region of England. Thirty of the permissions covering an area of 630 ha were for surface mineral workings. Only 2 permissions covering 50 ha were for spoil tips.

22.16 Permissions for opencast coal were the most extensive with aftercare conditions accounting for 44% of the total (300 ha). Sand and gravel (180 ha) and industrial sand (90 ha) permissions accounted for a further 27% and 13% respectively (Figure 22.7).

22.17 Agriculture was the most common afteruse proposed for sites with aftercare conditions accounting for 48% (330 ha) of the area. Amenity uses were proposed for 41% (280 ha) and forestry 11% (70 ha). The proportion proposed for agricultural use was well below the national average of 75%.

Reworking of Mineral working deposits
22.18 Three mpa's recorded sites permitted for the reworking of mineral working deposits in their area. Two mineral types, deep mined coal and clay/shale were involved (see Table J2 in Volume 2).

22.19 The total area permitted for such reworking was 140 ha, and involved 6 sites. One site of 45 ha was for colliery spoil being reworked under the GDO. The remaining five sites were worked under specific planning permissions, four colliery spoil sites (total of 80 ha) and 1 clay/shale site (10 ha).

Reclamation
22.20 There were 1430 ha of former mineral workings reclaimed in the North West region between 1982 and 1988, 7% of the total area reclaimed in England over that period.

22.21 Over 59% of this area was reclaimed by mineral planning permissions for the mineral (850 ha), the remaining 41% (580 ha) were reclaimed by "other means".

22.22 Former opencast coal and sand and gravel sites accounted for most of the land reclaimed by mineral planning permissions accounting for 38% (320 ha) and 30% (250 ha) respectively (Figure 22.8). Colliery spoil tip reclamation accounted for a further 17% (150 ha).

22.23 Almost two thirds (63%) of the land reclaimed by "other means" were former sand and gravel sites (370 ha). Clay/shale (100 ha) and sandstone (70 ha) accounted for a further 17% and 12% respectively.

22.24 Reclamation to amenity afteruses were the most common in the region, accounting for over 50% of the area reclaimed by mineral planning permissions (450 ha), and 73% of land reclaimed by "other means" (430 ha) (Figure 22.9). Agriculture accounted for 42% of land reclaimed by mineral planning permissions (350 ha) but for less than 5% of land reclaimed by "other means" (30 ha). Only 1 ha was reclaimed to forestry in the region.

22.25 Only 87% of land reclaimed by mineral planning permissions was considered to be satisfactory (740 ha). This was below the national average of 95%. In contrast 95% of land reclaimed by "other means" was considered satisfactory. The unsatisfactory reclamation of colliery spoil tips (60 ha) and sand and gravel sites (50 ha), were the main cause of the relatively poor reclamation record.

FIGURE 22.7 AREA OF MINERAL WORKINGS COVERED BY AFTERCARE CONDITIONS IN THE NORTH WEST REGION IN 1988 – MAIN MINERAL TYPES, AND PROPOSED ENDUSES.

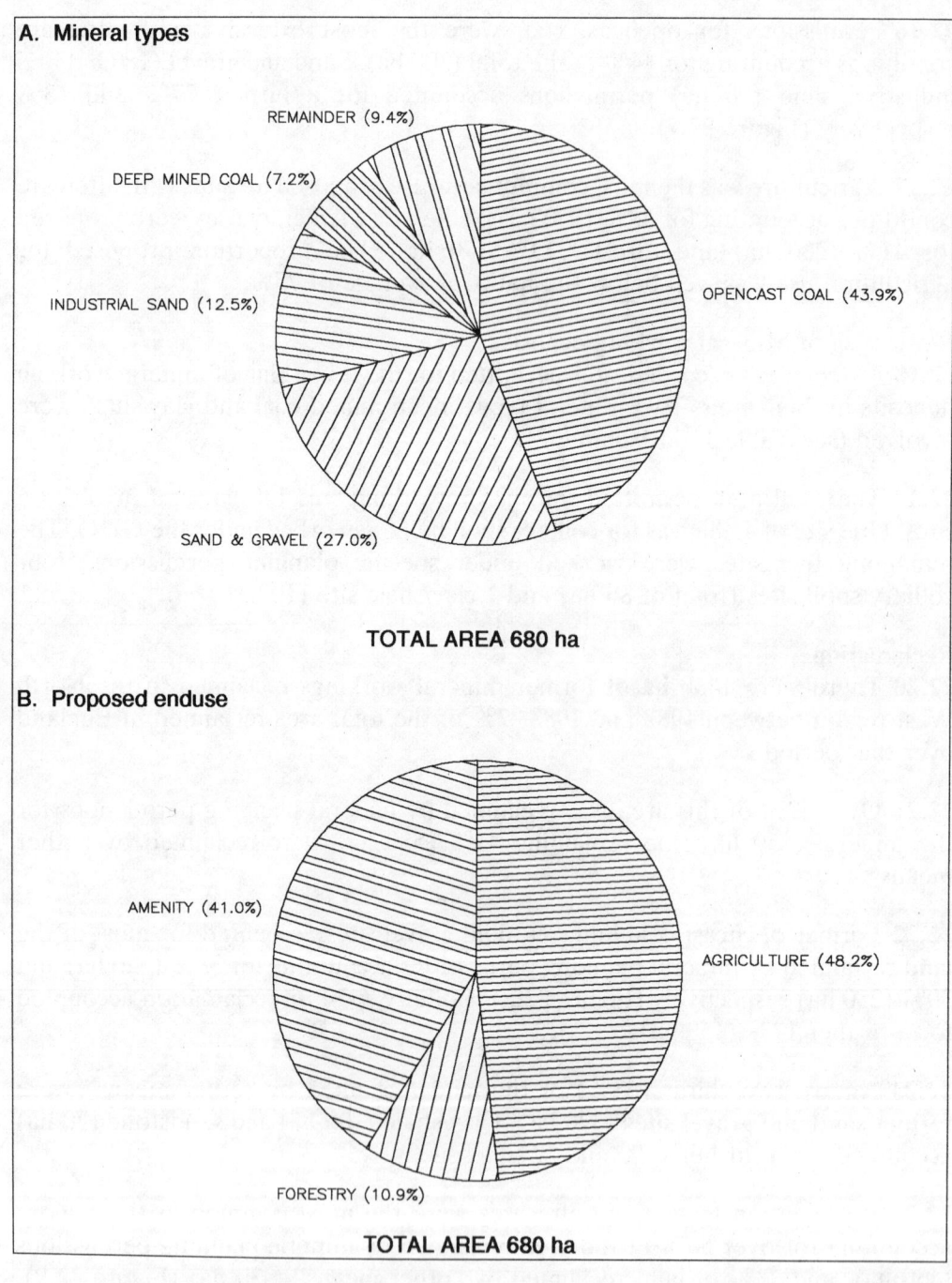

FIGURE 22.8 AREA OF LAND RECLAIMED BY MINERAL PLANNING PERMISSIONS (A) AND BY OTHER MEANS (B) IN THE NORTH WEST REGION BETWEEN 1982 AND 1988 – MAIN MINERAL TYPES.

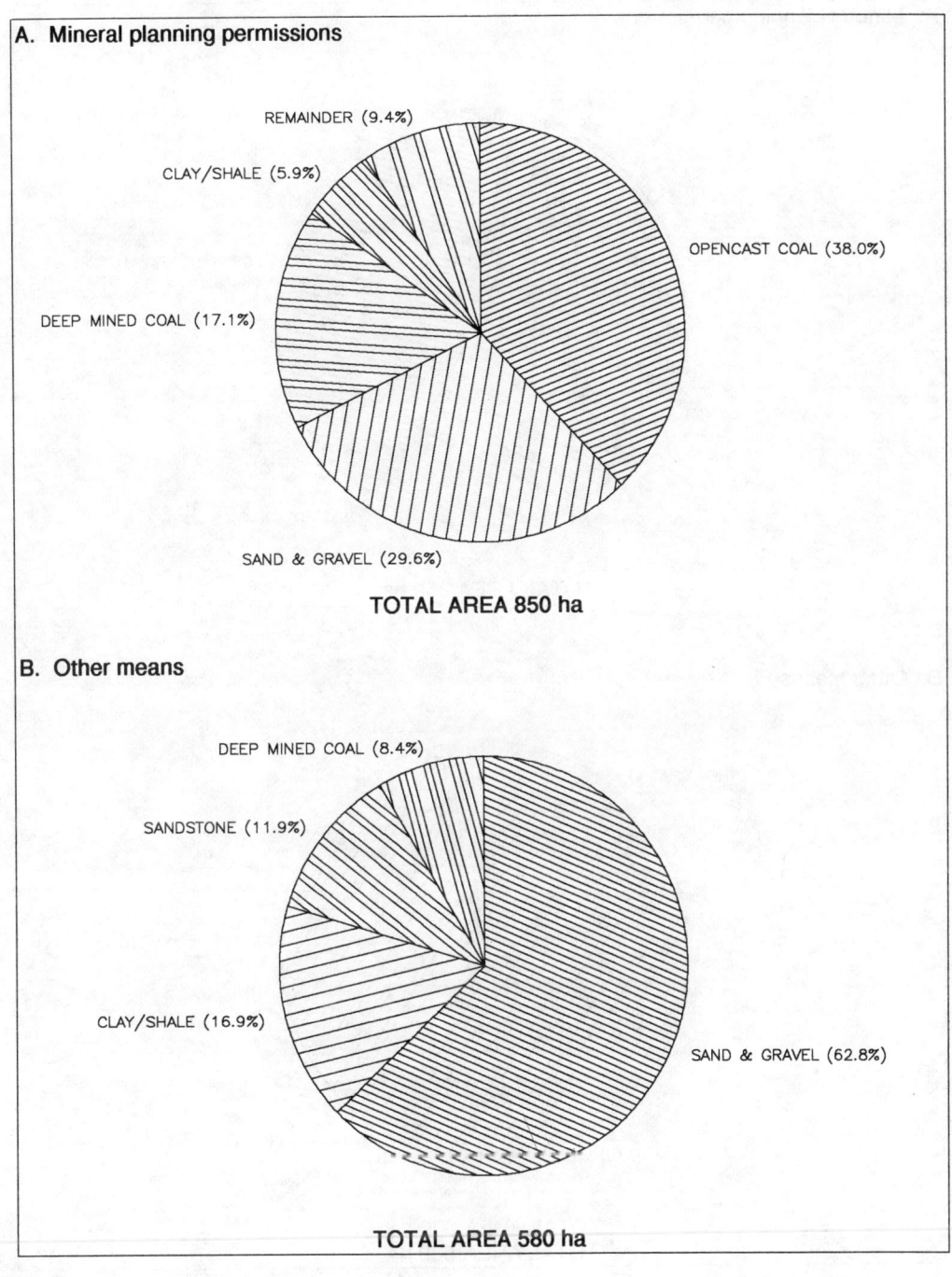

FIGURE 22.9 CHOSEN AFTERUSE OF LAND RECLAIMED BY MINERAL PLANNING PERMISSIONS (A) AND BY OTHER MEANS (B) IN THE NORTH WEST REGION BETWEEN 1982 AND 1988.

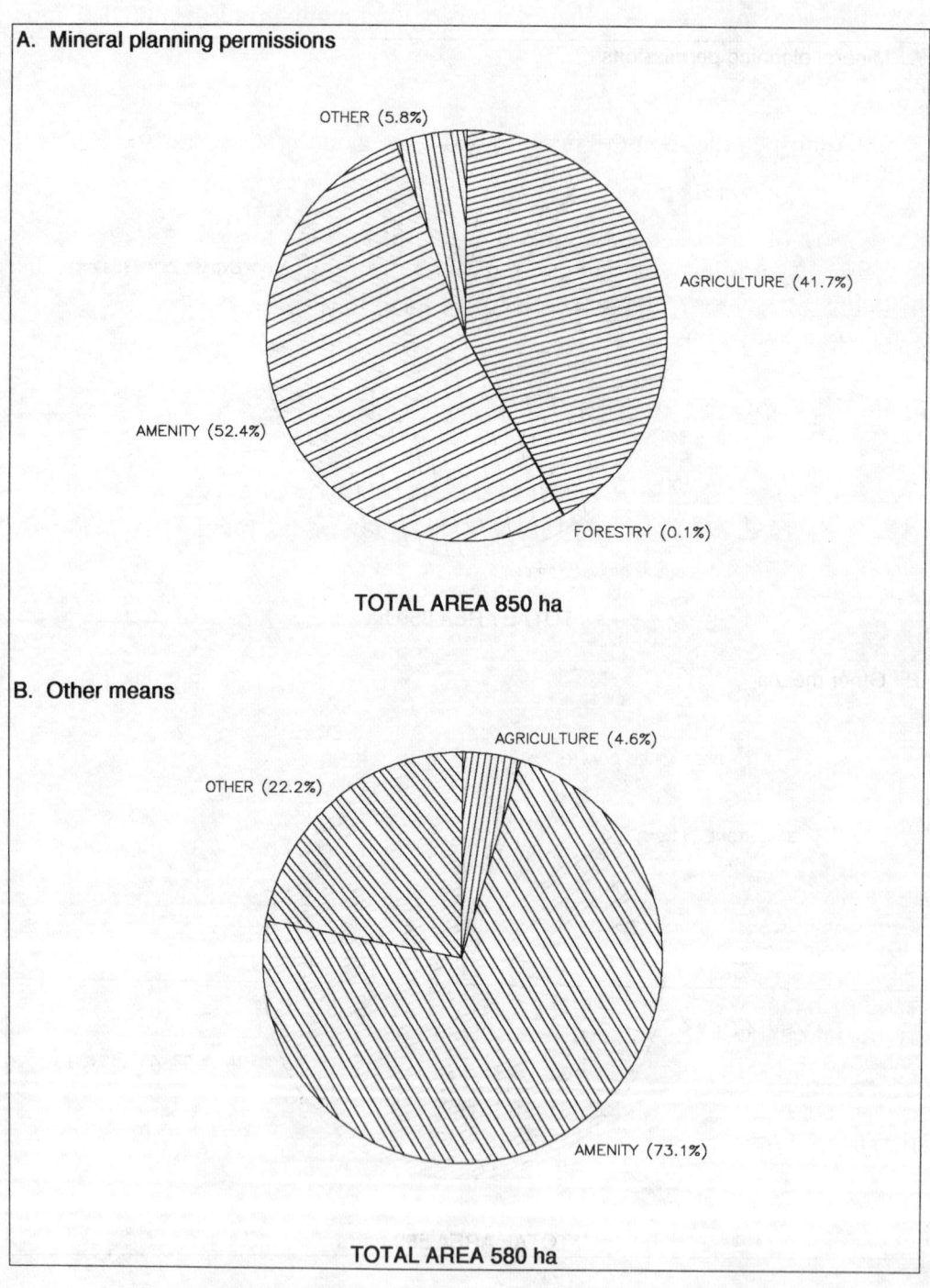

Number of sites

22.26 There were 375 sites associated with mineral working in the region. Most (78% or 292) were for surface mineral workings, 13% (48 sites) were for the surface disposal of mineral working deposits and 9% where underground mines (35 sites).

22.27 The region had the second smallest number of surface mineral workings in England, accounting for less than 7% of the total number in England. The most numerous were sandstone quarries (92), sand and gravel sites (76) and clay/shale pits (67) (Table 22.4).

22.28 Almost all the spoil tips permitted in the region (94%) were associated with coal mining (45 sites).

22.29 Most of the underground mines in the region were for coal (28) accounting for 80% of the total. Six mines were recorded for "other minerals" (mostly salt workings).

TABLE 22.4 NUMBER OF SITES FOR MINERAL WORKINGS IN THE NORTH WEST REGION IN 1988.

MINERAL	SURFACE MINERAL WORKINGS	SURFACE DISPOSAL OF MINERAL WORKING DEPOSITS	UNDERGROUND MINING
CLAY/SHALE	67	1	1
DEEP MINED COAL	0	45	28
OPENCAST COAL	15	0	0
LIMESTONE	9	0	0
OIL/GAS EXPLORATION	0	0	0
SAND & GRAVEL	76	0	0
INDUSTRIAL SAND	11	0	0
SANDSTONE	92	1	0
OTHER MINERALS	22	1	6
TOTAL	292	48	35

Land included in the Derelict Land Survey

22.30 There were 600 ha of land recorded in both this and the 1988 Derelict Land Survey. This was 14% of the total mineral dereliction recorded in England, and represented almost 12% of the current permitted area of surface workings and spoil tips in the region.

22.31 Surface mineral workings accounted for 60% (360 ha) of the derelict area, spoil tips 40% (240 ha).

22.32 More than half of the dereliction (56%), covering 330 ha, was a result of mineral workings with no provisions for reclamation. Most of this area (230 ha) was colliery spoil tips (Figure 22.10). The remaining 44% (270 ha) were areas associated with several different mineral types where provisions for reclamation were unfulfilled and are likely to remain so.

FIGURE 22.10 AREA OF LAND DAMAGED BY MINERAL WORKINGS IN THE NORTH WEST REGION AND INCLUDED IN THE 1988 DERELICT LAND SURVEY

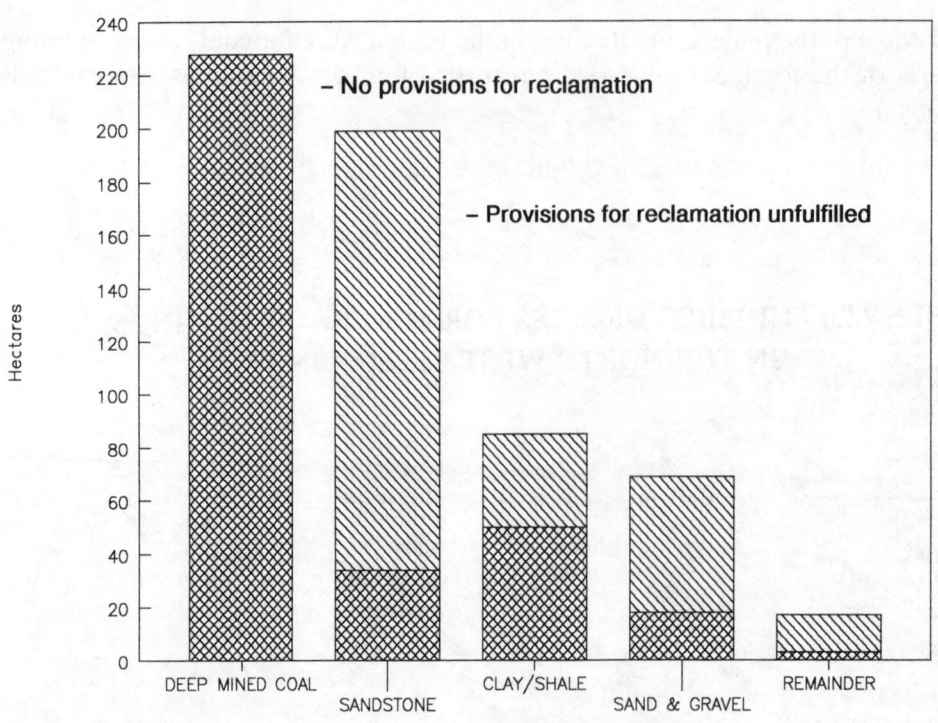

Reviews under the Town and County Planning Act 1971
22.33 None of the mineral planning authorities in the region had started their review procedures by 1 April 1988.

APPENDIX

Department of the Environment
Survey of Land for Mineral Working in England 1988

Contents

1. General Guidance Notes

2. Glossary

3. Tables for Completion (with detailed guidance notes) as below:

Surface Mineral Working and Surface Disposal of Mineral Working Deposits.

- Table 1A — Land covered by permissions with no conditions nor other arrangements for reclamation. (as at 1.4.88)

- Table 1B — Land damaged where all activity has now ceased and where conditions or other arrangements nominally providing for reclamation are unfulfilled and likely to remain so. (as at 1.4.88)

- Table 1C — Land covered by permissions which include conditions or other arrangements providing for reclamation. (as at 1.4.88)

- Table 1D — Land covered by permissions which include aftercare conditions. (as at 1.4.88)

- Table 2 — Permissions for underground mining. (as at 1.4.88)

- Table 3 — Mineral working deposits (other than 'stockpiles') being worked. (as at 1.4.88)

Reclamation of Mineral Workings

- Table 4A — Land damaged by surface mineral working or surface disposal of mineral working deposits and reclaimed under planning conditions or other arrangements for mineral working since 1 April 1982. (as at 1.4.88)

- Table 4B — Land damaged by surface mineral working or surface disposal of mineral working deposits and reclaimed under subsequent planning permissions for other development or the Derelict Land Grant programme since 1 April 1982. (as at 1.4.88)

- Table 5 — Number of sites and permissions relating to areas in tables 1A, 1B, 1C, 1D and 2. (as at 1.4.88)

- Table 6 — Reviews under Section 264A of the Town and Country Planning Act 1971. (as at 1.4.88)

Guidance notes

Number of returns

1. Because of the particular environmental significance of mineral working in National Parks and AONBs, we would like mineral planning authorities to fill in **separate returns** for AONBs and National Parks (or such parts of them as fall within the authority's boundaries). Thus each authority should return one set of forms for **each** AONB and NP — or part of them within its boundaries and one set of forms for the whole of the rest of the authority. The Peak District Board and The Lake District Board will, however, be making their own separate, returns.

Areas

2. Areas measured should include any margins which are to be left untouched around excavations or spoil tips (and which in some cases may be landscaped and/or used for the temporary storage of soil or overburden) and any areas designated for plant or other buildings. Where a number of planning permissions attach to a site, the site area will be the total area covered by the valid permissions. Any areas of sites which have been worked, but not reclaimed, without planning permission or under permissions which have lapsed, should also be included in the appropriate tables. For areas where permissions have lapsed and no working took place there should be no possibilty of future mineral working without a new permission, and therefore such areas should not be recorded. **All entries should be in whole hectares**. There should be no double counting of areas unless otherwise indicated.

Mineral types

3. Where 2 or more minerals are worked or deposited in the same site area, the entry should be made under the major mineral which promoted the application and development. There should be no double counting of areas, except (i) in the case of certain underground workings. Detailed guidance on this is given in the notes to Table 2. (ii) where there are clearly different permissions for working different surface minerals which partly overlap, such as where one permission has been completed and the land reclaimed, and the other has only partly been implemented. In this example the reclaimed site would be entered in Table 4A or 4B, and the other in Table 1 (A to D as appropriate).

Reclamation

4. The survey covers all areas affected by mineral workings and surface disposal of mineral working deposits except those which had been reclaimed at the time of the last 1982 survey or effectively reclaimed through natural revegetation. The identification of areas effectively reclaimed through natural revegetation must be an informed but subjective judgement; but should take account of the nature of the relevant permission and the likelihood of working or tipping activities restarting (if the latter, the areas would probably go in Table 1A).

Within this, the survey distinguishes areas active or where working has permanently ceased where no arrangements for reclamation including planning conditions have been made and areas where reclamation has not been effective because planning conditions or other arrangements for reclamation have been inadequate or unenforceable. We hope to see from the survey the extent of the problem of sites with reclamation conditions which you no longer consider adequate or which are, in your view, unenforceable. Inevitably there is an element of subjective judgement involved here, and in particular, what seems inadequate or unenforceable to a mineral planning authority may not agree with a strict legal interpretation of the situation. While leaving it to you to make your own judgements for the purposes of this survey, we would suggest that reclamation conditions should only be classed as inadequate or unenforceable where reasonable steps have already been taken to make progress with them or where an informed view can be made about conditions on sites where the reclamation stage has not yet been reached. It is suggested that authorities take into account the following factors: -

> (i) enforceability of conditions
>
> (ii) current restoration progress
>
> (iii) the record or attitude of the operator

Where there is no hope of enforcement against a landowner who is bankrupt, the area of land should be included in Table 1B. Include in Tables 1B and 1C as appropriate, sites where planning permission has expired but did nominally include reclamation conditions.

5. Where working ceased some time ago, the areas may be eligible for derelict land grant and may therefore be included in both the derelict land survey and here. When this occurs the areas which appear in both surveys should be recorded in Tables 1A and 1B as indicated by the notes for these tables.

6. Table 4A refers only to areas reclaimed under planning conditions or other arrangements for working minerals and seeks information about the after-uses of land. Table 4B refers to areas reclaimed by means other than mineral planning permissions, ie. planning permissions for other development or the Derelict Land Grant Programme. We are aware that there are areas of land which have already nominally been reclaimed under inadequate mineral or other planning conditions. Authorities, particularly those with large areas in this category, should indicate the scale of the problem by providing information within Tables 4A and 4B. Some guidance on what may constitute unsatisfactory reclamation is given in a footnote to these tables, but this will basically be a matter for the mineral planning authorities' judgement. Active sites whose current planning conditions are likely to lead to this situation should be included in Table 1C.

7. Tables 1A, 1B, 1C and 1D include all unreclaimed and unrevegetated land affected by surface mineral working or tipping or on which there are outstanding permissions still to be worked. Notes to the tables explain which type of land falls to which table. Surface activities or tipping associated with underground mining should also be included in the appropriate columns of Tables 1A, 1B, 1C and 1D.

Aftercare

8. For the first time, the 1988 survey asks - in Tables 1D and 5 (columns 4 and 5) - for information about the number and areas of sites subject to aftercare conditions. The tables distinguish between surface mineral working and surface disposal of mineral working deposits. Within these categories, Table 1D asks for areas to be broken down into agriculture, forestry and amenity afteruses. Further guidance on completion is given in the notes to the tables.

Mineral working deposits

9. Tables 1A to 1D ask for information on the surface disposal of mineral working deposits (see definition in the Glossary). These tables should include all such deposits (eg. spoil tips, tailings, silt lagoons) which are separately identifiable from areas of existing or future surface mineral workings or resulting from underground extraction of minerals. Mineral working deposits should be taken to include 'Stockpiles' for the purpose of completing all tables except for Table 3.

Site numbers

10. We appreciate that there is no clear definiton of a site; it can be the geographically unified area being worked, the area worked by a particular company, or the area covered by a planning permission. The main body of the survey does not refer, therefore, to sites but to areas. It would, however, be useful for us to have some overall picture of the number of mineral working or tipping sites throughout the country. Table 5 therefore asks for total numbers of sites for each return (AONB, National Park or 'other'). For the purposes of Table 5 of this survey consistency is necessary and a site should be taken as a geographically unified area, worked by one company. If authorities have difficulty in deciding what to count as one site, we would be happy to discuss the problem with them. A footnote to Table 5 refers to the avoidance of double counting where sites cross boundaries. For the purposes of reviewing mineral working sites, and for making orders under Part III of The Town and Country Planning Act 1971, and in completing Table 6 mineral planning authorities will have to have some working definition of a site for their own purposes, and there may well need to be constraints on the definition of a site.

Subsidence

11. Unlike the 1982 Minerals Survey, this survey will not cover instances of land damage caused by subsidence due to underground mining operations. Instead, this aspect will be included in the 1988 Derelict Land Survey.

Reviews

12. Table 6 asks for information about the progress and coverage of statutory reviews of mineral working sites. It also asks for details of action arising from reviews to be broken down by reference to the use of order making powers and of voluntary agreements entered into with mineral operators and owners.

Data correlation

13. There is a degree of correlation between several of the survey tables. Although the tables do not repeat information, some do analyse different components of planning or stages pertaining to individual sites. In consequence the presence of data on one table automatically requires the presence of other data on another table(s). The main examples are that an entry in Table 1D will also mean entries under either, or both of 1C and 4A and the need to check the numbers of sites entered in Table 5 with the relevant earlier tables.

Supplementary information

14. The Department is asking for purely numerical information in a standard format, but if authorities feel that accompanying maps at local or site level would be valuable, these will be appreciated. However, maps should only be sent to complement, not to replace, the standard statistical information. Similarly any additional written comments about the statistical information which the mineral planning authority wishes to add will be welcome.

Glossary

Mineral working deposit — Is defined in the Town and Country Planning (Minerals) Act 1981 as "any deposit of material remaining after minerals have been extracted from land or otherwise deriving from the carrying out of operations for the winning and working of minerals, in on or under the land".

Stockpile — Is defined in The Town and Country Planning General Development (Amendment) (No.2) Order 1985 as "a mineral working deposit consisting primarily of minerals which have been deposited for the purposes of their processing or sale".

Clay and Shale — Includes fireclay, ball clay and potters clay.

Vein Minerals — Includes tin, copper, lead, silver, zinc, haematite, iron ore, barytes, calcspar, fluorspar.

Restoration — The treatment of an area after operations for the winning and working of minerals have been completed by the spreading of any or all of the following - topsoil, subsoil, and soil making material (cf section 5 of the T & CP (Minerals) Act 1981).

Aftercare — The treatment of land following restoration (as defined above) to bring it up to the required standard for use for agriculture, forestry or amenity. Such treatment may include planting, cultivating, fertilising, watering, draining or otherwise treating the land.

Reclamation — Operations which are associated with the winning and working of minerals and which are designed to return the area to an acceptable condition whether for the resumption of the former land use or for a new use. Reclamation can include both restoration and aftercare as defined in the 1981 Act. However, reclamation can also include activitites which take place before, during and after mineral extraction, such as correct stripping and protection of soils, filling, shaping to agreed contours and creation of planned water areas. References in the survey tables to '**Reclamation Conditions**' should be interpreted likewise, since the respreading of soil ('restoration') is not an operation appropriate to all sites (eg. planned lagoon areas following mineral extraction). Planning conditions designed solely to achieve landscaping or screening of a site during mineral extraction or waste tipping should not be interpreted as 'reclamation conditions'.

Amenity use — Is a broad category of after-use which may include open grassland for informal recreation, basic preparation for more formal sports facilities, amenity woodland, lagoons for water recreation and the conservation of landscape and wildlife.

Land worked — Include all land where topsoil has been stripped prior to working, and all intervening stages (areas of current working, plant site, awaiting fill etc.) until the soil has been respread, or final landscaping of a lagoon has taken place. At that stage, the area should be classed as 'Reclaimed' and for the purpose of the survey included in Table 4.

Area not yet worked — Measured as surface land-take and is the surface area which has not been disturbed.

Land damaged — Is land justifying reclamation in the view of the mineral planning authority.

'Active' and 'Ceased' — A 'ceased' site is one that in general terms would be likely to conform with the definition in Section 51A of The Town and Country Planning Act 1971 ie. 'permanently ceased' with no substantial working having taken place for 2 years and where the mineral planning authority considers any resumption of working is unlikely. However, mineral planning authorities are not expected to have completed formal Section 51A procedures, except for Table 6 if it is relevant. An 'active' site is one that does not conform with the above definition.

'Other arrangements' — Arrangements providing for reclamation other than planning conditions, eg Section 52 agreements on land not covered by planning conditions, other voluntary agreements or arrangement.
(cf Tables 1A-C & 4A)

Department of the Environment

Survey of Land for Mineral Working in England 1988

* **Before completing the tables, please read the Guidance Notes**

* Mineral Planning Authority

* Name in **BLOCK CAPITALS** of officer making this return

  ☎ ext:

  Date   19

* Please tick one box and complete as appropriate. A separate return must be completed for each National Park or AONB occurring within the mineral planning authority's area.

  These tables relate to land situated in:

  ☐ National Park _____ (name)

  ☐ Area of Outstanding Natural Beauty (AONB) _____ (name)

  ☐ Neither of the above

  Tick one box only.

* Please return all forms even if you are making a nil return.

* All queries should be referred to Mr. B. Freeland ☎ 01 – 212 – 8489 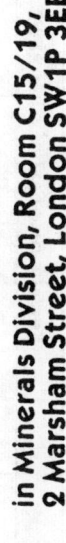 in Minerals Division, Room C15/19, 2 Marsham Street, London SW1P 3EB.

  or Mr. A. Wicks ☎ 01 – 212 – 3845

  The completed tables and this cover sheet should be returned to: Miss. A. Ward in Room C15/18 at the above address by 1st October 1988.

Table 1A

Land covered by permissions with no reclamation conditions nor any other arrangements for reclamation (as at 1.4.88)

To nearest hectare

For DOE use only: 1 ☐ 2 ☐ 3 ☐ 4 [1A] 5 [skip] — — See cover — —

| Mineral | Surface mineral working |||| Surface disposal of mineral working deposits ||||
| | Active || Ceased || Active || Ceased ||
	Total area worked (including plant) 1	Total area not yet worked 2	Total area worked and included in derelict land survey 3	Total area worked but not included in derelict land survey 4	Total area affected 5	Total area not yet affected 6	Total area affected and included in derelict land survey 7	Total area affected but not included in derelict land survey 8
1 Chalk								
2 China clay								
3 Clay/shale								
4 Coal (deep mined)								
5 Coal (open cast)								
6 Gypsum/Anhydrite								
7 Igneous Rock								
8 Ironstone								
9 Limestone/Dolomite								
10 Oil/Gas (exploration/appraisal)								
11 Oil/Gas (production)								
12 Sand and Gravel (construction)								
13 Sand (industrial/silica)								
14 Sandstone								
15 Slate								
16 Vein Minerals								
17 Other minerals								
18 Totals								

Notes

a. Include in this Table in columns 1,2,5 and 6 all such sites which are currently in use, sites where land remains to be worked or tipped on and partly worked and tipped sites which are not currently active but where these activities have been suspended and mineral reserves/suitable land for tipping remain within a valid consent.

b. Land covered by permissions yet to be implemented and having no reclamation conditions attached should be included in columns 2 and 6.

c. Include in columns 3,4,7 or 8 as appropriate, areas where surface mineral working or disposal of mineral deposits has now permanently ceased and where the land remains unreclaimed.

d. For surface mineral workings include **plant** etc. in columns 1,3 or 4 ie. as land worked, or, in the case of underground mining, in columns 5,7 or 8 ie. as land affected.

e. Include margins with the type of land, worked or not worked, that they surround.

f. Include in column 8 inactive sites for disposal of mineral working deposits which have naturally revegetated but which are still the subject of valid planning permissions (including permitted development under the GDO).

For DOE use only
1 ☐ 2 ☐ 3 ☐   4 [1B] 5 [SKIP]
- - - See cover - - -

Table 1B

Land damaged where all activity has ceased and where reclamation conditions or other arrangements nominally providing for reclamation are unfulfilled and likely to remain so. (as at 1.4.88)

To nearest hectare

	Mineral	Surface mineral working		Surface disposal of mineral working deposits	
		Total area included in derelict land survey 1	Total area not included in derelict land survey 2	Total area included in derelict land survey 3	Total area not included in derelict land survey 4
1	Chalk				
2	China clay				
3	Clay/shale				
4	Coal (deep mined)				
5	Coal (open cast)				
6	Gypsum/Anhydrite				
7	Igneous Rock				
8	Ironstone				
9	Limestone/Dolomite				
10	Oil/Gas (exploration/appraisal)				
11	Oil/Gas (production)				
12	Sand and Gravel (construction)				
13	Sand (industrial/silica)				
14	Sandstone				
15	Slate				
16	Vein Minerals				
17	Other minerals				
18	Totals				

Notes

a. Include in this Table mineral excavations worked under permissions requiring reclamation as and when fill becomes available but where filling is not now a realistic prospect.

b. Include sites where private arrangements for reclamation between landowner and the mineral operator have proved ineffectual.

c. Exclude sites where reclamation conditions have been fulfilled even where the land remains in an unsatisfactory condition (see Table 4). Damaged land where activity has ceased and there are no conditions outstanding may be eligible to be included in the derelict land survey if it is incapable of beneficial use without treatment.

d. Natural conditions (eg. high fells) may prevent reclamation. Exclude such sites here but where some reclamation arrangements have been made include them in Table 1C.

e. Exclude sites where working/disposal has ceased within the past five years and the efficiency of the reclamation conditions has not been adequately tested (see guidance notes para. 4).

f. Please indicate below if the information given is based on records going back less than 5 years.

Table 1C
Land covered by permissions which include reclamation conditions or other arrangements providing for reclamation (as at 1.4.88)

To nearest hectare

For DOE use only

| 1 | 2 | 3 | 4 | 1C | 5 skip |

- - - See cover - - -

	Mineral	Surface mineral working					Surface disposal of mineral working deposits			
		Satisfactory conditions or arrangements for reclamation		Unsatisfactory conditions		Total area dependent on imported fill and remaining to be filled	Satisfactory conditions or arrangements for reclamation		Unsatisfactory conditions	
		Total area worked but not yet reclaimed	Total area not yet worked	Total area worked but not yet reclaimed	Total area not yet worked		Total area affected but not yet reclaimed	Total area not yet affected	Total area affected but not yet reclaimed	Total area not yet affected
		1	2	3	4	5	6	7	8	9
1	Chalk									
2	China clay									
3	Clay/shale									
4	Coal (deep mined)									
5	Coal (open cast)									
6	Gypsum/Anhydrite									
7	Igneous Rock									
8	Ironstone									
9	Limestone/Dolomite									
10	Oil/Gas (exploration/appraisal)									
11	Oil/Gas (production)									
12	Sand and Gravel (construction)									
13	Sand (industrial/silica)									
14	Sandstone									
15	Slate									
16	Vein Minerals									
17	Other minerals									
18	Totals									

Notes

a. This Table distinguishes between sites with satisfactory reclamation conditions and those with potentially unsatisfactory conditions. (see guidance notes para 4).

b. Include in this Table all sites currently in use, areas where activities have ceased and reclamation is in progress up to the point where the soil has been replaced, where activities have been temporarily suspended, and inactive sites where there is no reason to doubt that reclamation will be complied with in due course.

c. Exclude land where working/disposal has ceased and reclamation conditions have been fulfilled, even where the land remains in an unsatisfactory condition. Where it is incapable of beneficial use, it may be included in the Derelict Land Survey.

d. Include in column 5 the total area to be filled with mineral working deposits which are not generated on site and controlled waste - ie. household, industrial or commercial waste as defined in Section 30 of the Control of Pollution Act 1974.

e. Include plant etc. in colums 1,3,6 and 8.

f. Include margins with the type of land, worked or not worked, that they surround.

g. Columns 1,2,3,4,6,7,8 and 9 together give the total area for this category. The figures in column 5 form part of these totals. This is recognised: the 'no double counting rule' for areas applies only to columns 1,2,3,4,6,7,8 and 9.

Table 1D
Land covered by permissions which include aftercare conditions (as at 1.4.88)

For DOE use only

| 1 | 2 | 3 | 4 | 1D | 5 | SKIP |

--- See cover ---

To nearest hectare

	Mineral	Surface mineral working			Surface disposal of mineral working deposits		
		Agriculture	Forestry	Amenity	Agriculture	Forestry	Amenity
		1	2	3	4	5	6
1	Chalk						
2	China clay						
3	Clay/shale						
4	Coal (deep mined)						
5	Coal (open cast)						
6	Gypsum/Anhydrite						
7	Igneous Rock						
8	Ironstone						
9	Limestone/Dolomite						
10	Oil/Gas (exploration/appraisal)						
11	Oil/Gas (production)						
12	Sand and Gravel (construction)						
13	Sand (industrial/silica)						
14	Sandstone						
15	Slate						
16	Vein Minerals						
17	Other minerals						
18	Totals						

Notes

a. Include in this Table all sites currently in use, sites with a valid planning permission where activity has not yet commenced, sites where activity has ceased and restoration/aftercare is in progress, where activities have been temporarily suspended and inactive sites where there is no reason to doubt that aftercare will be complied with in due course.

b. It is realised that these areas will also have been included under 'satisfactory conditions or arrangements for reclamation' in Table 1C. (columns 1, 2, 6 and 7) or in Table 4A since reclamation can include aftercare as defined in the glossary. Allowance will be made for this double counting in the presentation and analysis of the final survey information.

Table 2

Permissions for underground mining (as at 1.4.88)

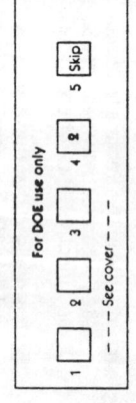

To nearest hectare

	Mineral	Total area 1	Extent of overlapping between underground workings 2
1	Clay/Shale		
2	Coal (under GDO)		
3	Coal (specific planning permissions)		
4	Gypsum/Anhydrite		
5	Ironstone		
6	Limestone/Dolomite		
7	Salt (including brine pumping)		
8	Slate		
9	Vein Minerals		
10	Other Minerals		
11	Totals		

Notes:

a. The areas in column 1 should be the projected extent of the permissions at the surface. This is normally quoted in the original planning application. Please asterisk particularly uncertain figures.

b. Give the maximum extent in column 1 where there are overlapping permissions given at different dates for mining of the same minerals at different depths.

c. If more than one mineral is mined at one site, return the site under the mineral which promoted the application or development (see paragraph 3 of guidance notes), unless there are separate permissions for mining totally different and geologically unrelated minerals (eg. coal from Carboniferous strata and Gypsum/anhydrite from overlaying Trias). In these cases make separate returns for each mineral and note the extent of overlapping as a total in column 2.

d. For vein minerals, please include the whole area of the permission within which the ore bodies lie, even if these areas are very large.

e. Where the area of the site is not given on the planning permission or working is taking place under the General Development Order, areas will be difficult to assess and particularly imprecise. Minerals officers will need to make a professional judgement of the likely area. This problem particularly applies to areas of permissions for underground coal mining. For this mineral category, therefore, separate entries are requested.

For DOE use only

| 1 | 2 | 3 | 4 | 5 skip |

--- See cover ---

Table 3

Mineral working deposits (other than 'stockpiles') being worked (as at 1.4.88)

Mineral	Under specific planning permission		Under GDO Class XXVII	
	No. of sites 1	Area (hectares) 2	No. of sites 3	Area (hectares) 4
1 Chalk				
2 China Clay				
3 Clay/Shale				
4 Coal (deep mined)				
5 Coal (open cast)				
6 Gypsum/Anhydrite				
7 Igneous Rock				
8 Ironstone				
9 Limestone/Dolomite				
10 Sand and Gravel (construction)				
11 Sand (industrial/silica)				
12 Sandstore				
13 Slate				
14 Vein Minerals				
15 Other Minerals				
16 Totals				

Notes:

a. These areas may also have been included in other tables, mainly Tables 1A to 1D. Were the information only to be recorded in this Table then important information about the nature of the reclamation conditions and other details provided about the areas under the headings in Tables 1A to 1D would be lost and comparisons with the 1982 Survey affected. Allowance will be made for this double counting in the analysis and presentation of the final survey information.

b. It is the primary source of the mineral working deposit which should be considered in completing this Table. The mineral type should therefore be interpreted as being the one involved in the original activity which gave rise to the deposit. For example, the reworking of colliery spoil tips should be recorded under coal (deep mined) and not coal (open cast).

Table 4A
Land damaged by surface mineral working or the surface disposal of mineral working deposits and reclaimed under planning conditions or other arrangements for mineral working since 1 April 1982. (as at 1.4.87) To nearest hectare

For DOE use only

1	2	3	4	4A	5
					skip

--- See cover ---

Mineral	Afteruse of reclaimed land							
	Agriculture		Forestry		Amenity		Other	
	Satisfactory	Unsatisfactory	Satisfactory	Unsatisfactory	Satisfactory	Unsatisfactory	Satisfactory	Unsatisfactory
	1	2	3	4	5	6	7	8
1 Chalk								
2 China clay								
3 Clay/shale								
4 Coal (deep mined)								
5 Coal (open cast)								
6 Gypsum/Anhydrite								
7 Igneous Rock								
8 Ironstone								
9 Limestone/Dolomite								
10 Oil/Gas (exploration/appraisal)								
11 Oil/Gas (production)								
12 Sand and Gravel (construction)								
13 Sand (industrial/silica)								
14 Sandstone								
15 Slate								
16 Vein Minerals								
17 Other minerals								
18 Totals								

Notes
a. In areas of progressive restoration, include areas which have been reclaimed as far as replacement of topsoil (see Glossary) as well as areas which have reached a formal aftercare or other land management stage.

b. Do not include land which has been naturally revegetated.

c. Only include an area as unsatisfactorily reclaimed if it still presents a problem for present and future use. Unsatisfactory reclamation may include, on filled sites, significant uneven settlement, continued emission of landfill gases or large rocks impeding cultivation. Other examples may be lack of soil, failure of plant growth in final landscaping, or water areas which are too shallow or irregular in depth.

d. Only include areas covered by planning conditions for mineral working including lagoons, grassland and woodlands for amenity and wildlife conservation. Exclude areas reclaimed under subsequent planning permissions for other development and land reclaimed under the Derelict Land Grant programme. These latter areas should be included in Table 4B.

e. See Glossary for the broad definition of amenity use (columns 5 and 6).

Table 4B
Land damaged by surface mineral working or the surface disposal of mineral working deposits and reclaimed under subsequent planning permissions for other development or The Derelict Land Grant Programme since 1 April 1982. (As at 1.4.88) To nearest hectare

For DOE use only

| 1 | 2 | 3 | 4 | 4B | 5 skip |

--- See cover ---

Mineral	Afteruse of reclaimed land							
	Agriculture		Forestry		Amenity		Other	
	Satisfactory	Unsatisfactory	Satisfactory	Unsatisfactory	Satisfactory	Unsatisfactory	Satisfactory	Unsatisfactory
	1	2	3	4	5	6	7	8
1 Chalk								
2 China clay								
3 Clay/shale								
4 Coal (deep mined)								
5 Coal (open cast)								
6 Gypsum/Anhydrite								
7 Igneous Rock								
8 Ironstone								
9 Limestone/Dolomite								
10 Oil/Gas (exploration/appraisal)								
11 Oil/Gas (production)								
12 Sand and Gravel (construction)								
13 Sand (industrial/silica)								
14 Sandstone								
15 Slate								
16 Vein Minerals								
17 Other minerals								
18 Totals								

Notes

a. In areas of progressive restoration, include areas which have been reclaimed as far as replacement of topsoil (see Glossary) as well as where vegetation has been planted and the after use commenced.

b. Do not include land which has been naturally revegetated.

c. Only include an area as unsatisfactorily reclaimed if it still presents a problem for present and future use. Unsatisfactory reclamation may include, on filled sites, significant uneven settlement, continued emission of landfill gases or large rocks impeding cultivation. Other examples may be lack of soil, failure of plant growth in final landscaping, or water areas which are too shallow or irregular in depth.

d. Only include areas reclaimed under subsequent planning permissions for other development and land reclaimed under The Derelict Land Grant Programme. Exclude areas covered by planning conditions for mineral working including lagoons, grassland and woodlands for amenity and wildlife conservation. These latter areas should be covered in Table 4A.

e. See Glossary for the broad definition of amenity use (columns 5 and 6).

Table 5
Number of sites and permissions relating to areas in Tables 1A, 1B, 1C, 1D and 2. (as at 1.4.88)

For DOE use only

1	2	3

4	5 skip

- - - See cover - - -

	Mineral	Surface mineral working (No. of sites) 1	Surface disposal of mineral working deposits (No. of sites) 2	Underground mining (No. of sites) 3	Permissions with aftercare conditions	
					Surface mineral working (No. of permissions) 4	Surface disposal of mineral working deposits (No. of permissions) 5
1	Chalk					
2	China clay					
3	Clay/shale					
4	Coal (deep mined)					
5	Coal (open cast)					
6	Gypsum/Anhydrite					
7	Igneous Rock					
8	Ironstone					
9	Limestone/Dolomite					
10	Oil/Gas (exploration/appraisal)					
11	Oil/Gas (production)					
12	Sand and Gravel (construction)					
13	Sand (industrial/silica)					
14	Sandstone					
15	Slate					
16	Vein Minerals					
17	Other minerals					
18	Totals					

Notes

a. Note that this Table seeks information about numbers of **permissions** with aftercare conditions (columns 4 and 5) as well as the number of **sites** involved in mineral working.

b. Site numbers entered in columns 1 – 3 should relate to the areas entered in Tables 1A – 1C and 2. Similarly, numbers of permissions entered in columns 4 and 5 should relate to the areas entered in columns 1 to 3 and 4 to 6 respectively of Table 1D.

c. Count each working site (columns 1 – 3) physically separate from another. Where a site has several planning permissions but is operated by one company, count as one site. Where a site has one planning permission but is operated by more than one company, count as one site per company. (see guidance notes para 10).

d. To avoid double counting when sites cross a National Park or AONB boundary, include the site in the National Park or AONB return but not in the 'other' return for your authority. If a site crosses an mpa border each authority should include it in their return. Contentious cases may be indicated separately (eg. in map form) if this would be helpful.

Table 6
Reviews under Section 264A of the Town and Country Planning Act 1971 (as at 1.4.88)

For DOE use only: 1 ☐ 2 ☐ 3 ☐ = 6 ☐
--- See cover ---

Date review started: D ☐ M ☐ Y ☐ = 5 ☐

	Mineral	No. of sites where working has commenced		No. of unimplemented permissions		Type of action for sites & unimplemented permissions					No. of voluntary agreements
		Reviewed	Identified for further action	Reviewed	Identified for further action	Section 45		No. of formal orders			
						Revocation	Modification	Section 51	Section 51A	Section 51B	
		1	2	3	4	5	6	7	8	9	10
1	Chalk										
2	China clay										
3	Clay/shale										
4	Coal (deep mined)										
5	Coal (open cast)										
6	Gypsum/Anhydrite										
7	Igneous Rock										
8	Ironstone										
9	Limestone/Dolomite										
10	Oil/Gas (exploration/appraisal)										
11	Oil/Gas (production)										
12	Sand and Gravel (construction)										
13	Sand (industrial/silica)										
14	Sandstone										
15	Slate										
16	Vein Minerals										
17	Other minerals										
18	Totals										

Notes
a. Columns 1 and 2 and 3 and 4 do not represent double counting. Columns 1 and 2 should be used to record information about sites where work has actually commenced. Columns 3 and 4 should record information about sites for which permission has been given but where work has not commenced.
b. The number of formal orders made for each mineral type should be recorded under the Section of the 1971 Act used in columns 5 to 9 as appropriate.
c. Include in column 10 the number of agreements made under Section 52 or other powers.

Printed in the United Kingdom for HMSO.
Dd. 293037, 6/91, C6, 3385/4, 5673, 136822.